T0091551

UNDERSTANDING ARTIFICIAL INTELLIGENCE

UNDERSTANDING ARTIFICIAL INTELLIGENCE

Nicolas Sabouret

Illustrations by Lizete De Assis

CRC Press

Taylor & Francis Group

Boca Raton London New York

CRC Press is an imprint of the
Taylor & Francis Group, an **informa** business

A CHAPMAN & HALL BOOK

Comprendre l'intelligence artificielle – published by Ellipses –
Copyright 2019, Édition Marketing S.A.

First Edition published 2021
by CRC Press
6000 Broken Sound Parkway NW, Suite 300, Boca Raton, FL 33487-2742

and by CRC Press
2 Park Square, Milton Park, Abingdon, Oxon, OX14 4RN

© 2021 Taylor & Francis Group, LLC

CRC Press is an imprint of Taylor & Francis Group, LLC

ISBN: 978-0-367-53136-2 (hbk)
ISBN: 978-0-367-52435-7 (pbk)
ISBN: 978-1-003-08062-6 (ebk)

Typeset in Times New Roman
by MPS Limited, Dehradun

To Chuck, with all my friendship.

Table of Contents

Introduction

Humans have never stopped inventing tools. During prehistoric times, we cut stones and made harpoons for hunting. At the dawn of agriculture, we invented pickaxes and sickles for farming, carts to transport heavy loads, and even tools to make other tools! Over time, they became more and more complex.

And then we invented machines.

The first were the cranes of antiquity. More than simple tools, machines transform energy to accomplish tasks that humans would otherwise have difficulty completing. The car, the washing machine, the lawn mower: we couldn't live without these inventions anymore.

Artificial intelligence is sometimes presented as a new revolution. It seems possible to provide machines with self-awareness, to make them capable of thinking for themselves, even to surpass us. This is a fascinating perspective, but it's also troubling. This is why each advancement in artificial intelligence gives rise to all sorts of fantasies.

Our ancestors certainly had the same reaction – a mix of fear and nervous enthusiasm – when the first weaving looms appeared during the Renaissance. These machines were capable of doing work that, up to that point, could only be done by humans. Suddenly, a human was replaced by a human creation. Some people considered it the "devil's machine," while others looked at it and saw a future where no one would have to work any longer. Clothing would make itself! History has shown that the truth is somewhere in between.

In 2016, many people were left in awe as AlphaGo, Google's artificial intelligence program, beat Lee Sedol at the game Go. The program even went so far as to make a move no human player would ever have thought of. We had built a machine capable of surpassing humans on their preferred terrain: strategic thought.

So what? Are we surprised if a car beats Usain Bolt in the 100-m dash? Humans have always tried to go beyond their limits. Artificial intelligence is simply a tool to do so. This is why we urgently need to demystify what AI means for us.

AI has very little to do with the way it's portrayed in science fiction novels. It is nothing like human intelligence, and there is no secret plan to replace us and enslave humanity. It is simply a tool, one of the best we have ever invented, but it has no will of its own.

AI is an incredible force that is changing the world we live in. But just as with any tool created by man, one must learn how to use it and guard against misuse. To do so, we must understand artificial intelligence. That is the goal of this book: I want to take you on a walk in the land of AI. I want to share my amazement at this extraordinary tool with you.

However, I also want to help you understand how AI works and what its limitations are.

What is Artificial Intelligence?

1

Understanding What a Computer, an Algorithm, a Program, and, in Particular, an Artificial Intelligence Program Are

What is artificial intelligence? Before we start debating whether machines could enslave humans and raise them on farms like cattle, perhaps we should ask ourselves what AI is made of. Let's be clear: artificial intelligence is not about making computers intelligent. Computers are still machines. They simply do what we ask of them, nothing more.

COMPUTER SCIENCE AND COMPUTERS

To understand what a computer is and isn't capable of, one must first understand what computer science is. Let's start there.

Computer science is the science of processing information.[1] It's about building, creating, and inventing machines that automatically process all kinds of information, from numbers to text, images, or video.

This started with the calculating machine. Here, the information consists of numbers and arithmetic operations. For example:

$$346 + 78 = ?$$

Then, as it was with prehistoric tools, there were advancements over time, and the information processed became more and more complex. First it was numbers, then words, then images, then sound. Today, we know how to make machines that listen to what we say to them (this is "the information") and turn it into a concrete action. For example, when you ask your iPhone: "Siri, tell me what time my doctor's appointment is," the computer is the machine that processes this information.

COMPUTERS AND ALGORITHMS

To process the information, the computer applies a method called an *algorithm*. Let's try to understand what this is about.

When you went to elementary school, you learned addition: you have to put the numbers in columns, with the digits correctly aligned. Then, you calculate the sum of the units. If there is a carried number, you make note of it and then you add the tens, and so on.

```
    1   1
    3   4   6
+       7   8
=   4   2   4
```

This method is an algorithm.

---

1   Indeed, in some languages, such as French and German, computer science is called "informatics," which has the same root as "information."

The algorithms are like cooking recipes for mathematicians: crack open the eggs, put them in the bowl, mix, pour in the frying pan, and so on. It's the same thing. Like writing an omelet recipe for a cookbook, you can write an algorithm to describe how to process information. For example, to do addition, we can learn addition algorithms and apply them.

When building a calculator, engineers turn these algorithms into a set of electronic wires. We obtain a machine capable, when provided with two numbers, of calculating and displaying the resulting sum. These three notions (the cooking recipe, the algorithm, and the electronic machine applying the algorithm) vary in complexity, but they are well understood: a cook knows how to write and follow a recipe; a computer scientist knows how to write an algorithm; an electrical engineer knows how to build a calculator.

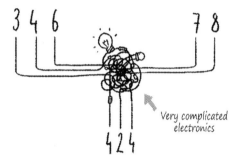

Very complicated electronics

ALGORITHMS AND COMPUTER SCIENCE

The originality of computer science is to think of the algorithm, itself, as information. Imagine it's possible to describe our addition recipe as numbers or some other symbols that a machine can interpret. And imagine that, instead of a calculator, we're building a slightly more sophisticated machine. When given two numbers and our addition algorithm, this machine is able to "decode" the algorithm to perform the operations it describes. What will happen?

The machine is going to do an addition, which is not very surprising. But then, one could use the exact same machine with a different algorithm, let's say a multiplication algorithm. And now we have a machine that can do both additions and multiplications, depending on which algorithm you give to it, at the same time.

I can sense the excitement reaching its climax. Doing additions and multiplications may not seem like anything extraordinary to you. However, this brilliant idea, which we owe to Charles Babbage (1791–1871), is where computers originated. A computer is a machine that processes data provided on a physical medium (for example a perforated card, a magnetic tape, a compact disc) by following a set of instructions written on a physical medium (the same medium as the data, usually): it's a machine that carries out algorithms.

THE ALL-PURPOSE MACHINE

In 1936, Alan Turing proposed a mathematical model of computation: the famous *Turing machines*.

A Turing machine consists of a strip of tape on which symbols can be written. To give you a better idea, imagine a 35 mm reel of film with small cells into which you can put a photo. With a Turing machine, however, we don't use photos. Instead, we use an *alphabet* – in other words, a list of symbols (for example 0 and 1, which are computer engineers' favorite symbols). In each cell, we can write only one symbol.

For the Turing machine to work, you need to give it a set of numbered instructions, as shown below.

```
Instruction 1267:

    Symbol 0 → Move tape one cell to the right,
              Go to instruction 3146

    Symbol 1 → Write 0,
              Move tape one cell to the left,
              Resume instruction 1267.
```

The Turing machine analyzes the symbol in the current cell and carries out the instruction.

In a way, this principle resembles choose-your-own-adventure books: *Make a note that you picked up a sword and go to page 37.* The comparison ends here. In contrast to the reader of a choose-your-own-adventure book, the machine does not choose to open the chest or go into the dragon's lair: it only does what the book's author has written on the page, and it does not make any decision on its own.

It follows exactly what is written in the algorithm.

Alan Turing showed that his "machines" could reproduce any algorithm, no matter how complicated. And, indeed, a computer works exactly like a Turing machine: it has a memory (equivalent to the Turing machine's "tape"), it reads symbols contained in memory cells, and it carries out specific instructions with the help of electronic wires. Thus, a computer, in theory, is capable of performing any algorithm.

PROGRAMS THAT MAKE PROGRAMS

Let's recap. A computer is a machine equipped with a *memory* on which two things are recorded: data (or, more generally, information, hence the word *information technology*) and an algorithm, coded in a particular language, which specifies how the data is to be processed. An algorithm written in a language that can be interpreted by a machine is called a *computer program*, and when the machine carries out what is described in the algorithm, we say that the computer is *running the program.*

As we can see with Turing machines, writing a program is a little more complex than simply saying "Put the numbers in columns and add them up." It's more like this:

```
Take the last digit of the first number.

Take the last digit of the second number.

Calculate the sum.

Write the last digit in the sum cell.

Write the preceding digits in the carrty cell.

Resume in the preceding column.
```

One must accurately describe, step by step, what the machine must do, using only the operations allowed by the little electronic wires. Writing algorithms in this manner is very limiting.

That's why computer engineers have invented languages and programs to interpret these languages. For example, we can ask the machine to transform the + symbol in the series of operations described above.

This makes programming much easier, as one can reuse already-written programs to write other, more complex ones – just like with prehistoric tools! Once you have the wheel, you can make wheelbarrows, and with enough time and energy you can even make a machine to make wheels.

AND WHERE DOES ARTIFICIAL INTELLIGENCE FIT IN ALL THIS?

Artificial intelligence consists of writing specific programs.

According to Minsky (1968), who helped found the discipline in the 1950s, AI is *"the building of computer programs which perform tasks which are, for the moment, performed in a more satisfactory way by humans because they require high level mental processes such as: perception learning, memory organization and critical reasoning."*

In other words, it's a matter of writing programs to perform information-processing tasks for which humans are, at first glance, more competent. Thus, we really ought to say "an AI program" and not "an AI."

Nowadays, there exist many AI programs capable of resolving information-processing tasks including playing chess, predicting tomorrow's weather, and answering the question "Who was the fifth president of the United States?" All of these things that can be accomplished by machines rely on methods and algorithms that come from artificial intelligence. In this way, there is nothing magical or intelligent about what an AI does: the machine applies the algorithm – an algorithm that was written by a human. If there is any intelligence, it comes from the programmer who gave the machine the right instructions.

A MACHINE THAT LEARNS?

Obviously, writing an AI program is no easy task: one must write instructions that produce a response that looks "intelligent," no matter the data provided. Rather than writing detailed instructions by hand, computer scientists often use sophisticated programs that extract the "right" answer automatically from some data source. As was the case for addition, the idea is simply to use the machine to reduce the burden of writing programs. As with addition, we try to use the machine to make writing programs easier. This principle is the key to an AI technique known as *machine learning.*

In fact, this name has caused a huge misunderstanding between computer science researchers and AI users. Obviously, this isn't a matter of letting programs loose in nature and expecting them to manage all on their own! We're simply using a property of computer science. The data processing (for example, how to do an addition) is described in a program that is itself a datum provided to the machine. A program that is also a datum can be modified or built by another program.

In this manner, you can write data-based programs capable of producing new AI programs.

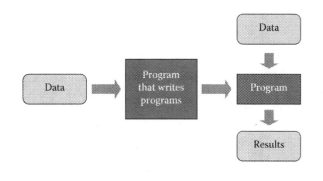

OBEY, I SAY!

Everything happens as if, like in the choose-your-own-adventure book, the author asked you to rewrite a page's content by replacing certain words with others, or by writing new instructions on blank pages at the end of the book. The information manipulated by the computer, whether data or the program itself, can be modified by the program.

However, this transformation only ever happens by obeying the algorithm's instructions. A program cannot produce *just any* program. If the initial program doesn't have good instructions, the resulting program will probably contain errors and be incapable of doing the most minimal "intelligent" processing. Let's return to our example of the choose-your-own-adventure book. If the author's instruction is to replace every "e" with "w," there is little chance you'll be able to continue the adventure after you've carried out the instruction on the page!

It is also important to remember that the operations carried out by your learning program, which produces the processing program, always depend on the data it has received. In our choose-your-own-adventure book, an instruction of this kind might be "Find the first letter of the first item you picked up." Thus, the modifications made depend on the data (here, the item you chose to pick up first). Obviously, if the data are poorly chosen, the resulting program won't be very effective either.

For this reason, an AI program using machine learning can only produce a "good" program if we give it good data and good instructions. None of this works all on its own!

ATTENTION: ONE PROGRAM CAN CONCEAL ANOTHER!

We are a long way from seeing a computer write a program on its own to resolve a problem it has not been designed for. It is too difficult. There are too many differences in the data sets, and too many possible rules to write for such a program to run correctly. An AI that could resolve any problem and make you a cup of coffee as a bonus simply does not exist because each problem requires a specific program and the data to match it.

To write AlphaGo, the champion Go program, the engineers at Google had to write an initial program to observe portions of Go games. Using these data, they wrote a second program capable of providing the best move in each situation. This required years of work! And they had to give the program good information so that it could *learn* from the data.

The result obtained is specific to Go: you can't just reuse it on another game. You could certainly adapt the analytical program and have it use different data to build another program capable of providing the best move in chess or checkers. It has been done by Google. But this requires the programmers to analyze the generation program, to modify it, to describe the rules of chess, and to explain how to analyze a chess board, which is, all in all, much different from a Go board. Your program cannot modify itself to produce something intelligent.

Would we expect a race car to be capable of beating eggs and washing laundry?

SO, WHAT IS AI?

To understand what AI is, we're going to look at some different methods used to write artificial intelligence programs.

All these methods rely on algorithms that are sometimes easy to understand but often difficult to put into practice. We will see how different tasks, each requiring the human ability to reason, has led researchers to invent methods to automate them. We will also see that, sometimes, these methods do not resemble at all what any relatively reasonable human would do in the same situation. Indeed, as Minsky (1968) once said, it's a matter of making *"computer programs which perform tasks which are performed in a more satisfactory way by humans."* It's not about making them perform a task as a human would do it.

Marvin Minsky's definition of AI is, thus, somewhat frustrating. From the start, he puts a limitation on AI. As soon as a task performed by a machine equals human beings in performing the given task, it is no longer AI. And it's kind of true: no one finds it extraordinary anymore that a GPS device shows the right way to go visit Grandma in Colorado just as well as if it had spent its whole childhood there. Yet it is indeed AI.

There are, of course, other definitions of artificial intelligence. The definition proposed in issue 96 of *La Recherche* magazine ("Research") in January 1979 is interesting. It states, *"Artificial intelligence seeks to equip computer systems with intellectual abilities comparable to those of humans."* Once again, this is fantasy: "equipping machines with intellectual abilities comparable to humans."

This is a very ambitious goal. It does make you dream a little, and that's why it's exciting. The real question, however, is to find out how far we can go in this direction.

REFERENCE

Minsky, M. (1968). *Semantic Information Processing*. The MIT Press, Cambridge, MA.

The Turing Test

Understanding That It Is Difficult to Measure the Intelligence of Computers

2

Before we develop any kind of artificial intelligence, it would be good if we define what intelligence is… and that's where things get complicated! While we know quite well how to recognize stupidity (all you have to do is look around you and see what people do wrong), it's harder to explain exactly what human intelligence is.

WHAT IS INTELLIGENCE?

The first idea that comes to mind is to define intelligence as the opposite of ignorance. Unfortunately, it's not that simple. If I ask you what year the city of Moscow was founded, you most likely won't know the answer. Yet no one could accuse you of not being intelligent just for that, even though the answer can be found in any encyclopedia. And you wouldn't say the encyclopedia is intelligent simply because it can tell you the date Moscow was founded. Intelligence doesn't consist of knowledge alone. You must also be able to use what you know.

A second idea would be to associate intelligence with the ability to answer difficult questions. For example, let's consider Georg Cantor, the mathematician who founded set theory in the 19th century. To deal with such in-depth mathematical problems, he was clearly very intelligent. However, one's ability in math is not enough to characterize intelligence. If I ask you what 26,534 × 347 is, it might take you some time to get the answer. However, a calculator will give you the result in an instant. You wouldn't say that the calculator is intelligent, not any more intelligent than the encyclopedia.

Who is the most intelligent?

These two examples demonstrate that while computers are particularly well equipped to handle anything requiring memory and calculation (they were designed for this purpose), this doesn't make them intelligent, because human intelligence consists of many other aspects. For example, we are able to reason by relying on our past experiences. This is what a doctor does when making a diagnosis, but this is also what each of us does when we drive a car, when we plan a vacation, or when we work. We make decisions in situations that would be impossible to describe accurately enough for a computer. When a baker tells us he's out of flour, we understand that he can't make any more bread. There's no need for him to explain.

We also know how to learn new skills at school or at work. We are able to use previous examples to form new concepts, create new ideas, and imagine new tools, using our ability to reason. Above all, we are able to communicate by using words, constructing sentences, and putting symbols together to exchange complex, often abstract, notions. When you tell your friends about your favorite novel, they understand the story without needing to read the exact words in the book. Thanks to our intelligence, we see the world in a new way every instant of our lives, and we are capable of conveying it to those around us.

All of this goes to show that if we want to compare a machine's abilities to human intelligence, we're going to have to try something more than just memory and math.

A TEST, BUT WHICH ONE?

To measure human intelligence, it is possible to use certain psychological tests. The most famous one is the IQ (or intelligence quotient) test, invented in the 1950s by William Stern. To determine your IQ, the psychologist relies on exercises adapted to your age and your education. These exercises evaluate your capacity for logical reasoning, spatial reasoning, memorization, and so on. The more quickly you respond with the right answer, the higher your IQ score.

Obviously, this kind of test only measures a small part of human intelligence. Researchers are increasingly interested in other aspects such as emotional intelligence (the ability to understand one's emotions and those of other people) or artistic creativity. Moreover, an IQ score is not an absolute value: it only allows you to compare yourself with the average score of people who have taken this test. If you have an IQ over 100, it means you do better at this test than the average person.

We could thus ask ourselves whether a computer has an IQ above or below 100. Unfortunately, adapting an IQ test, or any other test, to test a machine is a delicate operation. To start with, you would have to "translate" the questions into a language that the computer can understand – in other words, into zeros and ones. When you take an IQ test, a psychologist shows you pictures and asks you questions verbally. A human would have to "program" these questions into the computer.

What could we really evaluate with such a test? The machine's intelligence? The intelligence of the AI program that allows it to answer the questions? The way the data were translated into the computer's language? We mustn't forget that any program has to be written by a programmer using algorithms and data, as we saw in the previous chapter. So, what do we measure? The intelligence of the programmer who created and programmed the algorithm? The quality of the data given to the machine? Once again, we can see that none of this has anything to do with the machine's "intelligence." At the least, a good portion of this intelligence comes from a human.

AND THEN THERE WAS TURING!

In 1950, Alan Turing proposed a different kind of test to study the ability of machines. Instead of measuring intelligence, his idea was to simply differentiate between man and machine. We call this the Turing test, and today it is still considered as a reference when it comes to AI programs.

The principle is quite simple. It is inspired by a game that was played at social events, though it has long since gone out of style. Here you have it anyway, so you can spice up your birthdays and surprise parties. In this game, a man and a woman each go into a different room. The other guests, who remain in the living room, can "talk" with the man or the woman by sending notes to them through the servants. The guests cannot speak directly with the isolated persons, and they do not know who is in which room. They have to guess who is the man and who is the woman, and they know that both of them are trying to convince the guests that they are the woman.

In the Turing test, you have a human in a room and a computer with an artificial intelligence program in the other. You can communicate using a keyboard (to type your messages) and a screen (to read their responses). Thus, you have a keyboard and a screen to speak with the human, and another keyboard and screen to speak with the program, but you do not know which one is connected to the AI program imitating a human and which one is connected to the real human typing the responses. To complicate things a little more, all the responses are given within the same time interval. The speed of the answers isn't taken into consideration: you can only consider the content of the answers to distinguish between the human and the AI.

IT CHATS...

The Turing test is a challenge that still keeps many computer scientists busy today. It has given rise to the famous "chatbots," the artificial intelligence programs capable of speaking and answering just about any question one might ask them, from "What do you call cheese that is not yours?" to "Did you know you have beautiful eyes?" An international competition has even been created to recognize the best chatbots each year: the Loebner Prize. In 1990, Hugh Loebner promised $100,000 and a solid gold medal to the first programmer to develop a chatbot capable of passing the Turing test. As no one succeeded, an annual competition was started in 2006: chatbots are presented to judges who attempt to make them respond incorrectly as fast as possible. It is very amusing. If you want to give it a try, search for any chatbot program. There are plenty of them on the internet and in smartphone applications. You'll very quickly get the chatbot to say something ridiculous. But you might also be pleasantly surprised by the quality of some responses. That being said, know that the official Loebner Prize judges, who are experts in the subject, are capable of identifying a chatbot in fewer than five questions.

A CHATBOT NAMED ELIZA

The first chatbot was developed in 1966 by Joseph Weizenbaum. It is named ELIZA.

It operates according to some very basic principles. For one thing, it is equipped with a sentence catalog of classic topics (family, computers, health). It spots a topic's key words in the user's sentence (family, mom, dad, sister, etc.), which allows it to systematically respond to its interlocutor:

YOU: Do you have a mom?

ELIZA: Do you want us to talk about your family? And when it doesn't know how to respond, it sends the question back to the user:

YOU: What do you call cheese that's not yours?

ELIZA: Why are you asking me what do you call cheese that's not yours? It simply asks questions, letting the user provide the dialog.

Today, chatbots are much more evolved. ALICE, a program developed by Richard Wallace in the 1990s and a winner of the Loebner competition on several occasions, contains more than 50,000 rules and can answer anything you might wish to ask it. However, no one has succeeded yet in fooling the Loebner jury and winning the prize... and no AI program has successfully passed the Turing test either. On a regular basis, teams of computer engineers announce they've succeeded. But they never fully follow all of Turing's requirements.

IT'S CONTROVERSIAL TOO!

Actually, the Turing test is criticized as a method for evaluating "the intelligence" of an AI program. Don't forget that it was developed in 1950, at a time when computers weren't capable of processing images to detect cancer cells, or of driving a car on the highway! AI looked a lot different than it does today.

The first serious criticism of the Turing test came in 1980 from John Searle, a philosopher who has contributed much to artificial intelligence. To explain the Turing test's limitations, he proposed the "Chinese room" experiment. Imagine that a person who has no knowledge of Chinese is put in a room and isolated from the rest of the world. We give him rubber stamps with Chinese characters (50,000 characters, in fact!) and a manual with very precise instructions describing, for each possible question in Chinese, the answer to be given (also in Chinese). Our person in the room does not need to understand Chinese. In the manual, he finds the series of symbols corresponding to the question, and he uses the stamps to copy the response provided by the manual. He hasn't understood the question. He doesn't even know what he has responded, and yet a native Chinese speaker located outside the room would affirm without hesitation that the person in the room speaks Chinese.

The Chinese room experiment illustrates very well that intelligence is most certainly not limited to manipulating symbols. However, everything the Turing test measures relies entirely on symbols. And this is all rather normal because a computer is a machine that is designed to process information represented by symbols: zeros and ones.

Thus, the Turing test cannot measure a machine's intelligence, only its ability to process symbols like a human.

MY COMPUTER AND I ARE NOT ALIKE!

The error is to believe that we can judge a machine's intelligence by comparing its behavior to a human's. Turing only wanted to know if we could build a machine capable of thinking. This is why he proposed comparing a machine's behavior with a human's. But today we know that it's not that simple: all intelligent behaviors are not necessarily human. Think back to the multiplication of 26,534 × 347. You didn't even try to find the answer! By contrast, humans can perfectly adopt erroneous behaviors. For example, they make typos in their answers much more often than computers. This is also a way of recognizing them in the Turing test!

In fact, the Turing test is not really useful when developing an AI program. Turing never intended for his test to be used to measure the intelligence of computers. Let us not forget that a program is simply a series of calculations conceived for a specific purpose: choosing the next move in a game of chess, deciding whether to turn left at the next intersection, recognizing whether a picture contains a cat or an elephant, and so on. To test the intelligence of AI programs, we evaluate their performance based on the task to be performed (playing chess, finding a path, sorting photos). We don't try to see whether it's possible to talk with them about their philosophical viewpoint on the problem they're meant to resolve. The program is, however, much more efficient than a human.

This is an important difference. While the creation of artificial human beings is an interesting problem that requires writing AI programs, the opposite is not true. It is not necessary to make artificial humans to develop AI programs. You can certainly program AIs capable of resolving a given task brilliantly without resembling a human.

To explain this difference, computer scientists Stuart Russell and Peter Norvig, known in universities around the world for their book *Artificial Intelligence*, offer an amusing analogy with airplanes. Aeronautical engineers build planes according to aerodynamics. They do not try to make machines that fly exactly as pigeons so that they can fool other pigeons.

The strength of computers is their ability to process large amounts of data very quickly. Whereas a human relies on reasoning or on experience, a machine can instantly test billions of solutions, from the most absurd to the most relevant, or search through gigantic databases.

When you solve a Sudoku, it would never occur to you to try all the digits possible in each square, one by one. However, that's exactly how a computer operates: If I put 1 here... Nope, that doesn't work. OK, 2 then. Ah, maybe. OK, let's see, if put 1 in the square next to it... Nope, that didn't work either. OK, 2 then... and so on. Each of these operations only takes a few nanoseconds for a computer: it can do hundreds of millions of them per second! You cannot.

This fundamental difference between human intelligence and what we call artificial intelligence was summed up well by Edsger Dijkstra, a computer scientist who has an important place in the history of algorithms. He once said, *"The question of whether machines can think is about as relevant as the question of whether submarines can swim."* We build machines that do not think. In some ways, however, they give us the impression they are intelligent. They are not.

Why Is It So Difficult?

3

Understanding There Are Limitations to What a Machine Can Calculate

We must accept a sad truth: machines are not intelligent. At least, not in the same sense as when we say a person is intelligent. In a way, this is comforting for our ego: the thing that allows a machine to accomplish difficult tasks is the intelligence humans put into their algorithms. Well, then, what do AI algorithms look like?

One might think they are really special algorithms with forms or structures different from other programs, thus creating an illusion that machines are intelligent. Nothing of the sort. Like all algorithms, they look like a cooking recipe. And like all programs, they are applied, step by step, by a machine that basically operates like the reader of a choose-your-own-adventure book.

However, it takes years of research, by scientists who themselves have completed years of study, to develop each AI method. Why is it so difficult? It is important to keep in mind that there is no general technique for making AI, but rather many different methods and, therefore, many different algorithms. They don't look alike at all, but they often have something in common. They are all designed to get around the two primary limitations computers have: memory and processing capacity.

LIMITATIONS OF COMPUTERS

Computers are machines equipped with incredible memory and are able to make a calculation in less time than it takes to say it. However, like all machines, a computer can reach its limits.

Take the photocopier, for example. It's a machine that allows you to copy documents very quickly. That's what it's made for. To measure how well it performs, ask your grandparents what they had to do with carbon paper, or think about the monk scribes of the Middle Ages. Yet each photocopier can only produce a limited number of pages per minute, which is determined by its physical constraints.

For computers, it's the same. They compute very quickly; that's why they were invented. To give you an idea, a personal computer developed in the 2010s can do billions of additions per second. That's a lot (by contrast, I still recall you never finished the multiplication problem in chapter 2). And this number hasn't stopped increasing since the 1950s! But there comes a time when it isn't enough.

If you need to make a calculation that requires 10 billion operations, you'll have to wait a few seconds for the result. But if your program needs a trillion operations to solve the problem you've given to your computer, you'll have to wait 15 minutes. And if it needs 100 trillion operations, you'll have to wait an entire day!

A REAL HEADACHE

You're going to say to me: 100 trillion operations, that's impossible, it's too much! Don't be fooled: when we write algorithms, the number of operations becomes impressive really fast. Let's use a silly example: imagine we wanted to write a program that makes schedules for a school. It's a problem vice-principals spend a portion of their summer on every year. They would surely be very happy if a machine did it all for them. First, we'll start with a program that calculates all the possible schedule combinations for the students. Next, we'll see how to assign the teachers.

OK, we might as well admit it right off the bat, this isn't going to work. But we'll try anyway.

We have a school with 10 rooms for 15 classes (it's a small school with only five classes per grade, let's imagine). For each hour of class, there are around 3,000 possibilities for choosing the ten classes that can meet then. And for each eight-hour day of classes, that makes 6 billion billion billion possibilities (a 6 with 27 zeros). Yikes! If we have to consider them one by one, we'll never finish, even with a computer that can do a billion operations per second.

In computer science, it didn't take long to come up against this limitation on computing power. Actually, this question is at the core of computer program development. The difficulty for a computer scientist isn't just about finding an automated process to resolve a given problem. The process must also produce a result in a reasonable amount of time.

Alan Turing, whom we have already spoken of, was faced with this difficulty when he worked on deciphering German secret codes during World War II. At the time, it was already possible to build a machine that could try, one after another, all the possible codes the Germans might use to encrypt their messages using the Enigma machine. Combination after combination, the machine would compare the encoded word with a word from the message until it found the right one. In theory, this should work. But in practice, these incredible machines weren't fast enough to decipher the messages. They were capable of doing a job that required thousands and thousands of men, but it wasn't enough. All the Germans had to do was change the Enigma combination every 24 hours, and the code breakers had to start all over again. To "crack" the Enigma machine's codes, Alan Turing's team had to find a different way that didn't require so many operations.

COUNTING OPERATIONS

An entire branch of computer science research is dedicated to this computational limit. It is known as *complexity*. The notion of complexity is also very important in artificial intelligence.

To really understand this difficulty, it is important to know there are two types of complexities: the complexity of algorithms and the complexity of problems.

The complexity of an algorithm is simply the number of operations necessary for the algorithm to resolve the problem in question. For example, consider the "preparing an omelet" algorithm below:

```
For each guest:
        If the carton of eggs is empty:
                Open a new carton of eggs.
        Take an egg.
        Break open the egg on the bowl.
        Scramble.
        Pour in the frying pan.
        Repeat 60 times:
                Let cook five seconds.
                Scramble.
```

For each guest, this algorithm performs two operations (take an egg and break it open). Additionally, for every six guests, it performs an additional operation (open a carton). Finally, the eggs have to be scrambled, poured into the frying pan, and cooked, which takes 122 operations. If N is the number of guests, then the algorithm performs a total of $122 + (N / 6) + (2 \times N)$ operations. Voilà, this is its complexity.

As you can see, the complexity of an algorithm isn't just a number: it's a formula that depends on the data of the problem in question. In our example, the complexity depends on the number of guests: the more guests we have, the more operations we need. It's the same for a computer. For a given algorithm, the more data we give it, the more operations it will have to perform to resolve the problem.

Computer scientists calculate an algorithm's complexity based on the amount of data provided to the algorithm. Thus, the algorithm's complexity depends on the "size" of the problem – in other words, the number of zeros and ones we have to write in the computer's memory to describe the problem in question.

THIS IS GETTING COMPLEX

But that isn't all. Computer scientists deal with another complexity: the complexity of the *problem*.

This is defined as the minimum number of operations needed to resolve the problem with a computer. In other words, this corresponds to the complexity of the fastest algorithm capable of resolving the problem.

The algorithm isn't necessarily known: this is a theoretical complexity. *If* we knew how to write a super algorithm, the best of all possible algorithms to resolve this problem, this is what its complexity would be like. This doesn't mean, however, that we know how to write such an algorithm. Quite often, computer scientists have found other algorithms, with even higher complexities, and they rack their brains trying to find better ones.

In summary, when we give a problem to a group of computer scientists, they will first use mathematical proofs to determine the minimum number of operations needed to resolve the problem using a computer (the problem's theoretical complexity). Next, they're going to develop algorithms that will attempt to come near this complexity limit, if possible.

But where this gets really fun is when there are a whole bunch of problems for which the theoretical complexity is already way too high to hope to make an effective algorithm. In other words, even the best possible algorithm, assuming we know how to write it, couldn't solve the problem in a reasonable number of operations – not even with the computers of the future, which will be millions of times faster!

SQUARING THE CIRCLE?

There are a whole slew of problems that a computer will never be able to perfectly solve, no matter what happens. Computer scientists already know this. Yet almost all of the problems AI attempts to solve fall in this category: playing chess, recognizing a face, preparing a schedule, translating a statement into a foreign language, driving a car, and so on. The number of operations needed to find an *exact* solution to the problem is so large that it goes beyond what we can imagine in terms of computing power.

Indeed, artificial intelligence algorithms attempt to find a solution to these "impossible" problems. Obviously, the proposed solution will not be perfect (as we just saw, this is impossible!). The algorithm will incorrectly recognize a face from time to time, or it will make an error when playing chess or translating a sentence. But this algorithm will work in a "reasonable" time – in other words, with a lot fewer operations than what the theoretical limit requires for an exact solution.

Very often, making AI consists of developing programs that give us a not-so-bad solution in a reasonable amount of time, when we know the right solution is simply not obtainable.

Throughout the rest of this book, we are going to see a whole bunch of artificial intelligence methods that are not only not intelligent, they also don't necessarily provide a good solution to the problem in question. Admit it, AI's "intelligence" is starting to look doubtful!

That being said, however, you're going to see that it still works pretty well.

Lost in the Woods

4

Understanding a First Principles of AI Method – Exploration

Let us begin our journey into the land of AI with a simple problem humans are quite good at: "finding their way."

A SMALL WALK IN PARIS

Imagine that you come out of the Bréguet-Sabin subway stop right in the middle of Paris, and you need to go to Place des Vosges to visit the Victor Hugo House. As shown on the following map, you need to take Boulevard Richard-Lenoir toward Place de la Bastille for a hundred yards and then turn right onto Rue du Pasteur Wagner. Next, cross Boulevard Beaumarchais and take Rue du Pas-de-la-Mule until you reach Place des Vosges, the second street on the left. The Victor Hugo House is on this street.

Building this path is a task that humans know how to do by using their intelligence. Unfortunately, not everyone has the same ability for such a perilous mission. If you don't pay attention, you might very well end up in Place de la République.

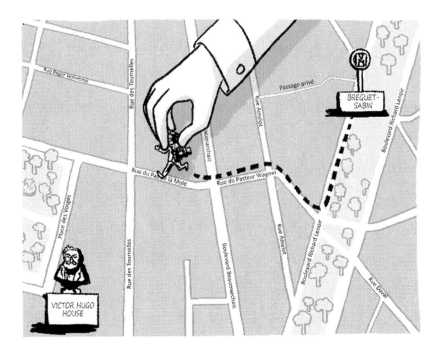

To get around this difficulty, nowadays we use a GPS, a device that allows us to determine our location on the earth's surface at any given moment. The GPS receives signals from twenty-eight satellites hovering a couple thousand miles over our heads. By measuring the time it takes each satellite to send a signal, it calculates the distance from these satellites and determines its position on the earth's surface.

HOW A GPS WORKS

To help you find your way, the GPS in your car or your telephone also has a map. This is an important tool that shows you all the places you can go and how they're connected. It is important to imagine what all this means in terms of data: each address, each portion of road, even if just a few yards long, each intersection ... everything must be represented on there, with GPS coordinates! Each possible position has a corresponding point on the map, and they are all connected to each other in a giant network.

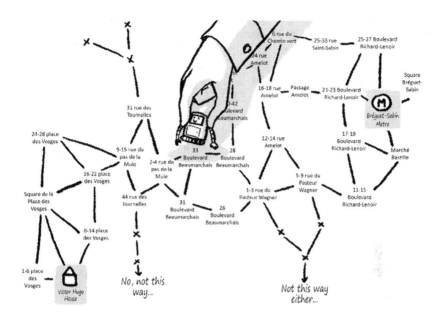

Thus, on the small portion of the map we used to go from the subway to the Victor Hugo House, there are already hundreds of points: the subway exit, the streets (cut up into as many little sections as necessary), all the intersections between these small portions of street, all the addresses of all the buildings on these portions of street ... and all these points are connected to each other.

Thanks to this map and the connections between the different points, the GPS can show how to go from the subway exit to Boulevard Richard-Lenoir, from the boulevard to the intersection with Rue du Pasteur Wagner, and so on.

FINDING THE PATH

A GPS map is a set of interconnected "points." In mathematics and computer science, this set is called a *graph*. It allows the artificial intelligence program to calculate the path to follow to go from one point on the graph to another.

The GPS knows what point you're located at (the subway exit). You've told it what point you want to go to (the Victor Hugo House). It must, then, calculate all the points you must pass through (all the small stretches of the boulevard, all the small stretches of Rue du Pas-de-la-Mule, etc.) to reach your destination.

From the GPS's point of view, your path is a series of points next to each other on the graph. In computer science, we say the GPS's AI program "finds a path on the graph." This path is then used by two other GPS components: the graphic interface, which allows us to view the path on the map, and the speech synthesizer (the little voice that gives you the information as you pass from one point to the next: "in 500 feet, turn left").

Let's consider the artificial intelligence program that determines the path.

IT IS MORE DIFFICULT THAN IT LOOKS

To develop a GPS pathfinder program, we have to write an algorithm that, when given a graph, a starting point, and an end point, is capable of *finding a path* on the graph. When we put it this way, we really get the sense that it's something a machine is capable of: we have all the necessary data. If you give a piece of paper with a graph on it to your six-year-old cousin and ask him to draw the path, he should be able to do it pretty easily.

But for a computer, this is where things get difficult. Remember what we saw about how they work: they are Turing machines that run operations on the "cells" on a strip of tape. Each cell corresponds to one specific piece of information for the machine, and it can only examine one cell at a time. Unlike us, the computer does not "see" the whole graph. It only sees one point at a time.

Put yourself in the machine's place and try to find a solution. You can only look at the starting point (the subway exit, in our example) and choose one of the neighboring points on the graph: Rue Bréguet, Rue Saint-Sabin, Boulevard Richard-Lenoir, Passage Amelot, apartment 21–23 on Boulevard Richard-Lenoir, and so on. You cannot examine them all at the same time: you have to choose one, save the ones you haven't looked at yet in your memory (in other Turing machine cases), and start over again at the next point.

If you choose Boulevard Richard-Lenoir, you can now look at a list of its neighbors. This list contains the "subway exit" point that you've already seen, but there are still all the others. You can add Rue du Pasteur Wagner to your list of points to look at. Next, you must choose another point and start again.

If you're fortunate, you'll end up making it to the Victor Hugo House with this method. But at no time will you know whether the point you've chosen is bringing you closer to, or farther away from, your goal. Again, you can only see your immediate neighbors on the graph.

What the program "sees"...

THE ADVENTURERS OF THE LOST GRAPH

It's quite unsettling, isn't it? The computer is like Theseus in the labyrinth: it can leave a trail behind it to know what path it has taken, it can remember each intersection it has come to, but it never knows where the corridor it has chosen will lead to!

This pathfinding problem on a graph is a classic artificial intelligence problem: computer scientists call it *state space exploration.*

Contrary to what the name suggests, it has nothing to do with sending a rocket into space with a flag of the United States. Not at all. In computer science, state space is the graph that defines all of the machine's possible states. Or, for a GPS, all the possible positions. State space exploration consists of walking through this space of possibilities, from one point to the next in search of a given point, all the while memorizing the path up to this point. It's Theseus in the labyrinth.

SO EASY!

Computer scientists do not consider this to be a "difficult" problem. If the graph is very large, the exploration will take a certain amount of time, but we know how to write some algorithms that can do it quite well. For example, when departing from the starting point, you can choose a random neighbor on the graph, then a neighbor of a neighbor, and so on. Basically, you randomly walk around until you find Place des Vosges. This algorithm is nice if you like to stroll about, but you may end up visiting the whole neighborhood before you actually come to the Hugo Victor House.

Another algorithm consists of looking at all of the starting point's neighbors first, then all of the neighbors' neighbors, and so on, until you come to the Victor Hugo House. You move in a spiral when you leave the subway. This algorithm is good if you don't want to miss anything in the neighborhood .... On the other hand, you won't exactly get to your goal right away.

However, these two algorithms have a very good complexity from a computer science standpoint: it is equal to the problem's theoretical complexity. This means that these two algorithms, while quite basic, are more or less what we can do best with a machine to find a path on a graph, even if we can see they pass through a ton of useless points.

Let us not forget that machines work "blindly" like Theseus in the labyrinth! We are guaranteed to find a solution with these two algorithms because they systematically explore the state space.

IT'S A WIN-WIN

If we look more closely, however, there is a big difference between these two algorithms. The second algorithm, moving in a spiral from the starting point, is going to study the points on the graph *in order based on their distance from the starting point*. That's why, when it achieves the objective, it will be certain to have calculated the shortest path (even though, in doing so, it had to go around in circles). This is clearly what we're looking for with a GPS. The first algorithm does not guarantee that the path it finds will be a "good" one. If it finds one, it may very well send you to Place de la Bastille, Place d'Italie, and Place de la République, with three changeovers in the subway, before you ever arrive at the Victor Hugo House.

Thus, the problem with the GPS isn't just exploring the state space, it's finding the *best* path possible on this darned graph.

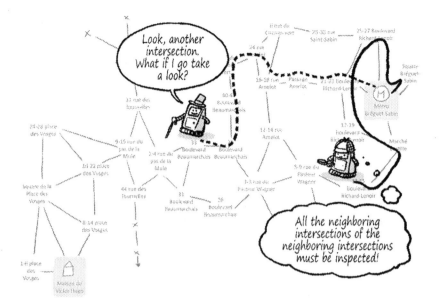

COMPUTER SCIENCE IN THREE QUESTIONS

This example illustrates a fundamental problem in computer science. When we create an algorithm, we always have to ask three questions.

First, does the algorithm address the problem in question? This is essential. Here, both algorithms find a solution, if there is one. No worries about that.

Second, does the algorithm achieve the result in a reasonable amount of time? This is the complexity we already discussed. Our two algorithms are more or less the same in terms of complexity, and they're actually very good because they reach the theoretical complexity limit.

Third, is the solution provided by the algorithm of "good quality"? For our pathfinding, a "good" solution is one that doesn't make us visit all of Paris. If we're clearly looking for the shortest path here, we better not be strolling about.

A CLEVER ALGORITHM

We owe the first effective algorithm for finding the shortest path on the graph to Edsger Dijkstra. This algorithm uses the spiral idea while considering the *distances* between the points on the graph (for example, the length of a section of street or of an alley).

On the three computer science questions, Dijkstra's algorithm scores 10 out of 10. First, it finds a path if there is one: we can prove it mathematically. Second, it does it in a reasonable amount of time: the algorithm's complexity is close to the problem's complexity. Third, it is proven that the path obtained is always the shortest of all the possible paths. If there is another path, it is undoubtedly longer.

Yet this algorithm is not an AI algorithm. First and foremost, it's an effective algorithm for exploring any kind of state space.

AI IS HERE!

To begin making AI, one must consider the data shown on the graph and the computing time.

With a GPS, the data are the points on the graph – that is, the different places where you can go. To show the Paris neighborhood we used in our example, we need a graph containing about a hundred points, perhaps more. For a small city, you'll quickly need about ten thousand points.

Dijkstra's algorithm, like most algorithms, requires more and more computation as the graph gets larger: its complexity depends on the size of the inputs, as we saw in the previous chapter.

Specifically, the complexity of Dijkstra's algorithm is the number of points on the graph squared. If a graph contains 10 points, we'll need around 100 operations (10×10) to calculate the best path. For 100 points, we'll need 10,000 operations (100×100), and so on. The greater the number of points, the greater the number of operations.

At this rate, we quickly reach the limit a computer can compute in a reasonable amount of time. And that's where artificial intelligence comes in. We need to find an "intelligent" state space exploration algorithm so we won't have to wait an eternity for the GPS to tell us what route to take.

A HEAD IN THE STARS

The artificial intelligence algorithm that can resolve this problem is called "A star" (or "A*" for those in the know). It was invented by Peter Hart, Nils Nilsson, and Bertram Raphael in 1968.

The idea behind this algorithm is simple: What takes time in Dijkstra's algorithm is ensuring that, at all times, the path produced will be the shortest path possible. What we need is an algorithm that, without exactly wandering around randomly, doesn't necessarily have to examine everything like Dijkstra's algorithm does.

This algorithm will not always give the best solutions. But we can hope that, if it doesn't choose the points completely at random, it will provide a, "good" solution. Naturally, we will have taken a bit of a stroll, but it will be in the right direction.

This is the principle of artificial intelligence.

THE BEST IS THE ENEMY OF THE GOOD

The first thing to understand about AI algorithms is that they *do not aim to obtain an exact solution to the problem posed.*

When a cooking recipe says "add 1.5 ounces of butter," you don't necessarily take out your scale and weigh your butter to the ounce. You take an eight-ounce stick of butter, cut off what looks like three-eighths, and that's it. It's a lot faster! By doing it this way, you are consciously forgoing an exact solution, but you know that your cake won't necessarily be that bad. It just won't be "perfect."

This is the idea behind artificial intelligence. When it is not possible to compute the best solution in a low enough number of operations, we must content ourselves with a faster, "approximate" solution. We have to give up on exactness to gain computing time for the machine.

ROUGHLY SPEAKING, IT'S THAT WAY!

In artificial intelligence, we call this a *heuristic.* It's a somewhat highbrow word for a method or algorithm that provides a solution that isn't necessary the best. However, if we're lucky, it isn't too bad either, and we get it in a reasonable amount of time. It's like cutting the butter without using the scale.

The A* algorithm uses a heuristic to explore the state space. This heuristic consists of going in the right direction as much as possible, while staying on the right path as much as possible. This is also what you do when you find your way on a map: you know where your destination is, and you look how to get there with your finger by tracing your finger in a mostly straight line from your starting point.

Likewise, when you walk along the street, you head in what you believe is the "right direction." If you come upon an obstacle, you go around it, or you retrace your steps. In general, if you have a good estimate of your initial direction, this method is quite effective.

This is exactly what our A* algorithm is going to do. Instead of choosing, at every step, the point closest to the starting point that it hasn't looked at yet, as Dijkstra does, it chooses a compromise between the distance from the starting point (to remain close to the shortest path) and the distance to the destination point as the crow flies (to come as close as possible to the goal).

FOLLOW THE CROWS

Computer scientists call the distance as the crow flies (straight line distance) a *heuristic function*. It's a rule of computation that gives a "not so bad" estimate of the distance to arrive at each point.

This function must be easy to calculate: it is going to be used each time the algorithm encounters a new point as it explores the state space. If it takes three minutes to calculate the heuristic function's output, we'll never finish! With the GPS, this works out well: the straight line distance is very easy to evaluate because you know each point's coordinates on the graph.

But the heuristic function must also give the best possible estimate of the actual distance to the destination point – that is, the distance you'll obtain when you take the path on foot, as opposed to flying over buildings. You might have some bad luck: there's construction work on Rue du Pas-de-la-Mule and you have to take a huge detour on Rue Saint-Gilles. But, in general, this seems like a good path to follow.

Actually, with the GPS, the straight line distance is an excellent heuristic. It is possible to mathematically prove that the path provided by the A* algorithm is clearly the shortest path. Computer scientists say that this heuristic is *admissible*.

This is not always the case. There are many computer science problems for which one must find a path on a graph. The A* algorithm can be used with different heuristics, each one corresponding to a specific problem. The special thing about this algorithm is precisely how generic it is. Even if the heuristic function provided to the machine to resolve the problem is not admissible (in other words, it doesn't guarantee it will find the shortest path on the graph), the A* algorithm will find a solution rather quickly. It won't necessarily be the best, but with a bit of luck it won't be too bad.

THERE'S A TRICK!

The GPS example allows us to understand that inside each AI algorithm there is inevitably a trick provided by a human: the heuristic.

Resorting to heuristics stems from the fact that it is impossible for a computer to guarantee an exact solution to the problem posed in a reasonable amount of time. AI algorithms always calculate an approximation, and it's the heuristic function (and, of course, a well-chosen algorithm) that allows us to obtain solutions that are "not too bad." The quality of the solution will give people the sense that the machine is more or less intelligent.

With the A* algorithm, the programmer writes the heuristic as a function – for example, the straight line distance for a GPS. In other AI programs, as we'll see throughout this book, other heuristics will be considered. In this way, computer scientists have created an illusion of intelligence by giving the algorithm instructions that allow it to find a good solution. For its part, the computer simply applies the heuristic. Undoubtedly, this is very rewarding for the computer scientists who make AI.

Winning at Chess

5

Understanding How to Build Heuristics

We now have in our suitcase all the equipment we need to make our journey to the land of AI. A Turing machine, some algorithms, a bit of complexity, some graphs, and two or three heuristics: voilà, the computer scientist's complete survival kit.

To understand artificial intelligence, let's take a stroll through computer science history.

A SHORT HISTORY OF AI

Everything began in 1843, when Ada Lovelace wrote the first algorithm for Charles Babbage's analytical engine, a precursor to the computer. In a way, this is the prehistory of computer science. The first programmable machines didn't quite exist yet: Jacquard's loom, invented in 1801, was already using perforated cards to guide needles, but it was purely a matter of controlling mechanical commands. The first machines to use computation and programming only appeared at the end of the 19th century, during the US census. The first entirely electronic computer, the ENIAC, was invented in 1945, a

100 years after Ada Lovelace's algorithm. It is where we entered year zero in computer science... and everything would go much faster after this.

In 1950, just five years after the arrival of the ENIAC, Alan Turing used the term "artificial intelligence" for the first time. This new concept was a massive hit, and in just a few years numerous AI algorithms were proposed.

In the 1960s, computer science became widespread. The computer quickly became an essential tool in many industries, given its ability to store and process data. The A* algorithm we discussed in the previous chapter dates back to this time period. In the 1970s, computers took over offices: employees could directly input and consult company data. Everyone was talking and thinking about AI!

And yet it would take almost 20 more years for computers to beat humans at chess. What happened?

During this time, AI hit a rough patch. After the first successes in the 1960s, society was full of enthusiasm. People imagined machines would soon be capable of surpassing humans in every activity. They thought we would be able to automatically translate Shakespeare to Russian and replace general practitioners with computers that could make a diagnosis based on a patient's information.

The reality check was brutal. Investors who had had a lot of hope for this promising field realized that AI only provided good results in a few limited domains. And just as before, a computer scientist had to systematically give the machine very specific algorithms and data. Without them, it was no more intelligent than a jar of pickles.

It was very disappointing. At a time when we knew how to calculate a rocket's trajectory to go to the moon, we couldn't even program a robot capable of crossing the street without getting run over. The fallout with AI was quite simply on par with hopes it had generated.

Unfortunately, it was all just a big misunderstanding between computer scientists on one hand, who were all excited by the prospect of having the first intelligent tasks completed by machines, and the general public on the other hand, who got carried away by what they saw in science fiction.

Artificial intelligence paid a heavy price. From 1980 to 2010, approximately, only a few businesses agreed to invest in these technologies. Instead, most preferred to stick with exact algorithms they could manage and monitor easily.

Feng-Hsiung Hsu and Murray Campbell, two young IBM engineers, began their career at this dire time. They set a goal that was both ambitious for AI and exciting for the general public: to build a computer program capable of defeating chess's grand champions.

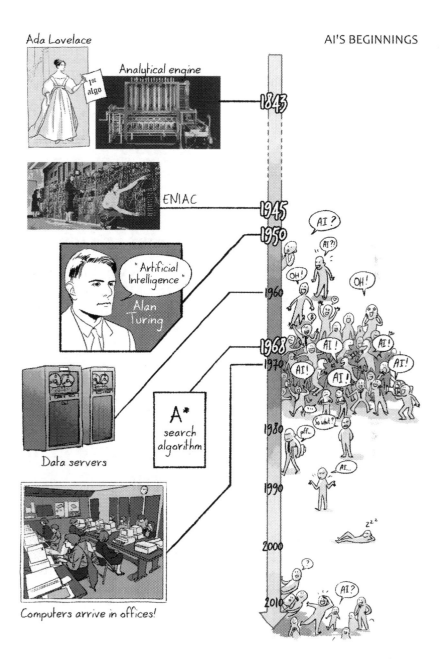

Ada Lovelace

Analytical engine

1st algo

AI'S BEGINNINGS

ENIAC

1843

1945

1950

AI ?

AI ?!

OH!

OH!

"Artificial Intelligence"

Alan Turing

1960

1968

AI!

AI!

1970

AI!

AI!

AI!

Data servers

A*
search
algorithm

1980

pff...

So wut?

1990

AI...

2000

zzz

2010

?

AI?

Computers arrive in offices!

A NEW CHALLENGE

Playing chess requires intelligence, without a doubt. It takes several hours to understand how each piece moves and years of training to play well. Chess champions are attentive observers and dedicate many hours of study to all the different tactics.

Building a machine capable of winning at chess is, therefore, a real challenge. It is not possible to "program" all the tactics: there are too many situations to consider, and too many interactions between the game pieces and positions to be able to write rules for a computer. You have to go about it a different way.

The "Deep Blue" project developed by IBM's research department was up to the task. Hsu and Campbell were hired right after they graduated from Carnegie-Mellon University, where they had worked on something similar as part of a student project. IBM's research director was very shrewd. He let them pursue this project with computing resources that were rare at that time. In the 1980s, IBM was in fact building supercomputers that could process information much faster than any ordinary computer. The very first IBM chess player, a machine called Deep Thought,[1] could evaluate 3,000 configurations per second!

However, it lost to Garry Kasparov in 1989. Finally in 1997, Deep Blue, a machine composed of 256 processors capable of evaluating 200 billion configurations in three minutes, beat the undefeated world champion... and just barely!

AN OLD ALGORITHM

Quite surprisingly, the algorithm Hsu and Campbell used to win at chess wasn't new. It existed before computers were even invented! This algorithm was proposed by the mathematician John von Neumann, another big name in computer science. Von Neumann is known, among other things, for developing the electric circuit that would later be called a computer.

---

1   The name Deep Thought comes from the supercomputer in Douglas Adams's humorous novel *The Hitchhiker's Guide to the Galaxy*. In the book, the computer computes the "Answer to the Ultimate Question of Life, the Universe and Everything," no less! According to the author, the answer is... 42!

Long before this, in 1928, von Neumann took an interest in "combinatorial games." These are two-player games with perfect information and without randomness, of which chess is but one example: there are two players who see the same thing (complete information) and there is no randomness, contrary to snakes and ladders, for example. Chess and checkers are probably the two best-known combinatorial games in Western culture. But have you ever played Reversi, where players flip over discs? And the African game known as Mancala, where each player puts seeds in small pits to "capture" opponents' seeds? John von Neumann demonstrated that, in all of these games, there is an algorithm for determining the best move. This is known as the *minimax theorem.*

Note that we cannot just simply apply this algorithm. This is what makes it such a subtle thing. Consider as an example the board game Hex, a Parker game that came out in the 1950s. In this game, each player tries to link two opposing sides with hexagonal pieces, while blocking the opponent. The minimax theorem proves there is an algorithm that will allow us to calculate the best move. John Nash, the inventor of Hex and future Nobel Prize winner in economics in 1994, even succeeded in proving there is a strategy that allows the first player to win every time! Automatically computing this strategy (in other words, with the help of an algorithm), however, is such a complex problem that it is completely impossible for a computer to do it in a day, even if it were billions of times more powerful than today's most powerful computers.

In other words, it has been shown that the first player in Hex will inevitably win if he chooses the right moves (his opponent cannot stop him from winning)... but it is impossible to calculate what the moves are!

FROM THEORY TO PRACTICE

The algorithm that allows computers to win at chess against the best players in the world takes its inspiration from the method defined by the minimax theorem. Naturally, it's called *the minimax algorithm*. The idea is relatively simple: attempt all the moves possible, then all the opponent's possible moves (for each possible move), then all the possible responses, and so on until one of the two players wins.

This builds a kind of genealogical tree of possible game situations starting from the initial situation. In this tree, the root is the initial state of the game (no player has made a move), and each move corresponds to a branch leading to the state of the game once the move has been made, as shown in the following drawing.

*The "min-max" principle applied to tic-tac-toe:
building a genealogical tree.*

The idea behind the minimax algorithm is to use this genealogical tree to choose the best move. Starting at the "final" situations at the top of the tree (in other words, the endings of the games), we move downward toward the root.

To begin, we transform all of this into numbers because computers are machines that manipulate numbers. Then, we assign a score to each final situation. For example, in Mancala, the score could be the number of seeds obtained minus the opponent's seeds. In chess, the score could be 0 if you lose, 1 if you draw, and 2 if you win.

Imagine now that you are at the last move of the game. If it's not your turn, the logical choice for your opponent is to make the winning move. Put another way, your opponent is going to make a move that achieves the least optimal final situation for you (unless you're playing someone totally stupid). Thus, the other player's goal is to achieve the *minimum* score (from your point of view). Conversely, if it's your move, you want to choose the move leads you to the *maximum* score. Therefore, the score of a situation when there is only one move left is the final situation's minimum (or maximum) score.

And for the move right before it? It's the same, except the roles are reversed! You must choose the move that leads to the minimum (if it's your opponent's turn) or the maximum (if it's your turn) of what we just calculated for the final move. The score of each situation two moves from the end is, thus, the maximum of the minimum (or the minimum of the maximum: this depends on the player).

Thus, von Neumann's algorithm starts at the game's final moves at the top of the tree and moves "downward" toward the root by alternating the minimums and the maximums. Hence the algorithm's name: the move you must choose at the end is the maximum of the minimums of the maximums of the minimums of the maximums of the minimums. In sum, it's minimax!

The "min-max" principle applied to tic-tac-toe:
calculating the score of each board.

This is somewhat hard to imagine because we aren't machines. For a computer, calculating these maximums and minimums is very easy, just like exploring the state space we saw in the previous chapter. The program consists of less than ten lines of code!

In all of this there isn't necessarily any artificial intelligence. The reason for this is quite simple: when the minimax algorithm was invented in the late 1920s, a machine's computing power wasn't relevant. Computers didn't even exist yet!

ANOTHER LIMIT TO COMPUTERS?

To apply the von Neumann method directly, one must make a genealogical tree of all the possible games. This task varies in difficulty depending on the game in question.

Let's start with tic-tac-toe. Von Neumann's method can be applied directly. If we consider the symmetries and the fact that the players stop playing as soon as one of them has won, there are 26,830 possible games in tic-tac-toe. For a machine, examining these 27,000 configurations and calculating the maximums and minimums is actually a piece of cake! With several billion operations per second, your brain would take longer to read the result than the machine would to calculate the best move! There is no complexity problem here, so there's no need to resort to artificial intelligence. The von Neumann method works.

CHECKMATE, VON NEUMANN!

Now, let's take a look at chess, which is, you'll agree, slightly more interesting than tic-tac-toe. Not all the pieces can move over all the squares, and each kind of piece is subject to different rules. For this reason, calculating the number of configurations is more complicated than tic-tac-toe. According to the mathematician Claude Shannon, there are around 10^{120} possible chess games – in other words, 10^{120} different games to examine to choose the best move. To write this, we use the digit "one" followed by 120 zeros:

1,000,000,000,000,000,000,000,000,000,000,000,000,000,000,000,
000,000,000,000,000,000,000,000,000,000,000,000,000,000,000,
000,000,000,000,000,000

To give you a better idea, this is billions of billions of billions... of billions of times more than the number of atoms in the entire universe. It is impossible to imagine that a machine will someday be able to examine all of these games in a reasonable amount of time. Currently, computers are capable of going up to the first 10 zeros; that means there are still 110 more after that! Even if we assume the computing power of computers will continue to double every two years, as suggested by Moore's law,[2] we will have to wait 732 years before we are able to make a computer capable of examining all the possible games in under a second.

To beat Garry Kasparov, an artificial intelligence algorithm was needed. As we saw in the previous chapter, this means two things. First, forget about always winning: this is the *heuristic method*. Our computer will know how to play chess, but it will make mistakes: that's one of the rules of artificial intelligence. Second, we have to ask humans which calculations to make in this heuristic so we can reduce the number of mistakes.

THE RETURN OF MINIMAX

When adapted to artificial intelligence, the minimax algorithm takes the principle of the initial algorithm but adds two fundamental ideas. The first is to consider only a subpart of the tree; the whole is too large to be computed by a machine. Instead, the computer stops once it reaches a certain number of "reasonable" moves. In other words, since it is impossible to calculate all the games, we'll just look at what happens for the first eight to ten moves.

Even this makes for a lot of games to look at. To give you an idea, imagine each player has ten possible moves for each turn (there are generally many more). On the first move, made by white, ten possibilities. When it's black's turn, again ten possibilities. So, that's 100 (10×10) possibilities for a two-move game. On the next move, for white, again ten possibilities, and so on. On the tenth move, there are already ten thousand (10^{10}) different games to consider. A computer even from the early decades of the 2000s cannot go much beyond that.

With 10 moves, it's highly unlikely the game will be over. Actually, it will just be getting started. Nevertheless, it is possible to assign a score to each of these "game beginnings." This is the second idea introduced in the minimax algorithm: using a *heuristic function* provided by a human.

---

2 In 1975, the engineer and computer scientist Gordon Moore made a hypothesis that the number of transistors on a microprocessor doubles every two years. This assertion gave life to Moore's law, which in plain language is "computer processing power doubles every two years." This very imprecise assumption has become a target to aim for in the computer science industry, and it has pretty much been the norm since the 1970s.

A HEURISTIC TO WIN AT CHESS

If we look at a chessboard during a game in progress, it may not be easy to say who is going to win. But we can, nevertheless, say two or three interesting things. If you've already lost your queen, a rook, and two knights, the game is not off to a good start. In contrast, occupying the middle of the board early in the game is generally considered to be a good tactical advantage.

Chess players are able to assign scores to these different situations. When children learn to play chess, they learn the values of each piece: a queen is worth 10, a rook 5, the knight or the bishop 3, and so on. Using this idea, we can assign a score to the chessboard. Additionally, we can award a bonus if a player is occupying the center of the board, or a penalty if a player's pieces are under threat. This doesn't allow us to say for sure who is going to win, but it gives us an initial estimate.

Concretely, this means that, even after only ten turns, we can calculate a score for each chess game. In computer science, we program this score as a heuristic function.

From here, the principle is the same as with von Neumann's algorithm: the computer uses the score to calculate minimums and maximums. It's that simple! As we descend this initial genealogical tree from the tree's "leaves" (corresponding to the games of eight to ten moves) toward the roots (corresponding to the game's current situation), the algorithm is able to tell which move corresponds to the maximum of minimums of the maximums of the minimums, and so on. Thus, this is likely the best move at the moment.

Next, it's your opponent's turn. Here, the algorithm starts over based on the new situation: it calculates all the eight- to ten-move games in advance and applies the minimax method to choose the move that looks best. And so on and so forth.

EASIER SAID THAN DONE

With this algorithm, you can write a program that plays chess pretty well – except for a few key details.

For starters, the minimax algorithm can be improved by "cutting" the tree branches we already know won't provide a better result. This method is known as the *alpha-beta pruning* method. John McCarthy, one of the most famous researchers in artificial intelligence, proposed this pruning idea at a

Dartmouth workshop in 1956. This is not a new heuristic, because the cut branches are "clearly" bad according to the minimax theorem. Thus, in practice we don't lose anything. On the other hand, this pruning allows us to work with bigger trees and gain a few moves for your game estimate. Instead of only considering the next 10 moves, the computer can go up to 15 or 16 moves!

After this, to win against the world chess champion, you need to write a good heuristic function. This means you need to determine the value of the threatened pieces, the captured pieces, the different strategic positions, and so on. Nevertheless, all this requires a certain expertise in chess and quite a few attempts before you find the right values.

Finally, you need to have a supercomputer like Deep Blue and a whole team of engineers to work on it for a decade! A small consideration, of course.

A DISAPPOINTING VICTORY

The Deep Blue victory in the late 1990s did not give AI the renewed momentum we might have expected. The general public and specialists alike were not convinced by the quality of Deep Blue's playing. And for good reason! Some of the computer's moves were terrible. The algorithm does not plan everything, and, unlike humans, it has no strategic knowledge.

It wasn't exactly a resounding victory either: some matches ended in draws, and the decisive match appears to have been more the result of a mistake by Kasparov, disconcerted by an unlikely move by Deep Blue, than by strategic and tactical qualities worthy of chess's best masters!

Without knowing everything about the minimax algorithm, people easily understood that the IBM computer simply applied a crude method and won by brute force. This computing power, though not actual physical force, gives pretty much the same impression as if Hulk Hogan (a 275-pound heavyweight wrestling champion) won a judo competition against an Olympic champion who competes in the under-105-pound weight class. Something doesn't seem right.

But that's exactly how AI works. And not just in games and GPS. Remember: this is about allowing machines to perform *tasks that are, for the moment, performed in a more satisfactory way by humans.* The method isn't important, it's the result that matters.

Nothing says the techniques used have to resemble what a human would do. Quite the contrary, the power of computers is their ability to process large amounts of data very fast. Whereas a human will rely on reasoning and experience, a machine will test a billion solutions, from the most absurd to the most relevant, or search through giant databases in the same amount of time. It doesn't think, it computes!

So, this is somewhat disappointing. However, it would be absurd to expect machines to do things "like humans." They are not physically capable of it.

The Journey Continues

6

Understanding That One Graph Can Conceal Another

Unlike humans, machines do not "think." That's why it's difficult to make true artificial intelligence.

ON THE INTELLIGENCE OF MACHINES

As humans, we resolve problems by using our experience and our ability to reason. To develop AI algorithms, on the other hand, we have to "think AI." All we have is a Turing machine, a strip of tape with small cells containing data and instructions to go from one cell to the next. Therefore, we have to find a technique that works with this tool and its strengths. This means computing power, not reasoning.

To help themselves with this task, artificial intelligence researchers can take inspiration from nature, which often does things very well. You may already have heard of *genetic algorithms*, artificial *neural networks*, or *ant colony algorithms*. Behind these peculiar names are AI methods inspired, in one form or another, by nature.

A VERY PARTICULAR TRAVELER

To understand these different methods, let's consider an amusing problem known as the *traveling salesman* problem. Imagine you are a sales representative for some company – a company that makes air conditioners, let's say. Your job is to present your new models in several large cities throughout France. Naturally, you want to make the trip as efficient as possible by spending the least possible amount of time on the road.

When you put it that way, the problem looks similar to our GPS story. However, it is very different.

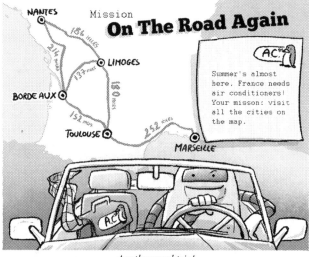

Another road trip!

A HEALTHY WALK

Naturally, it is also a graph problem. If we represent each city as a point on a graph and connect each point to all the others with "lines" showing the time required to go from one city to the next, we indeed get a graph that we'll have to travel across.

But this isn't about finding a path from one point to another. To start, all the points are connected to each other. So the path is a piece of cake! What we need is to find the shortest path that passes through *all* of the points on the graph.

This problem is very well known in computer science... and it is impossible to resolve it efficiently! It is nothing like the GPS problem or Dijkstra's algorithm: that's child's play compared to this! To give you an idea, if you try to find an exact solution to the traveling salesman problem for a graph with 20 points, the best algorithms possible will take a hundred years to give you an answer with a modern computer. With Dijkstra's algorithm, you get an answer in under a second for a graph with 1,000 points.

A REALLY DIFFICULT PROBLEM...

The traveling salesman problem has many applications in computer science, but that's not all. In the "door-to-door" category, in addition to urban furniture and sonic vacuum sales, it can also be used by an power company to determine the route the meter reader should take, by a pizza company for customer deliveries, by the post office to deliver mail as quickly as possible, by the city to pick up garbage, and so on.

The solution to this problem is also needed to optimize the computational costs for computer graphics cards, to test electronic circuits, or to control the movements of industrial machinery. The economic implications are huge!

Very early on, computer scientists sought to provide a "good" solution to the traveling salesman problem in a reasonable amount of time, in the absence of an exact solution.

THE GREEDY ALGORITHM

Over the course of time, several AI algorithms have been proposed. The first of them has a very original name: the *greedy* algorithm. This AI method uses a simple heuristic: at each stage, the traveler chooses the city that is closest to where he is (and which he hasn't visited yet, of course).

In computing terms, we just choose the one that's the closest distance away from among the remaining unvisited points. We can write the algorithm as follows:

Let C be the set of all the non-visited points in the graph.
Let p be the closest point in C to the current position.

Go to the point and remove it from C.
If C is empty, you are finished. If not, resume.

This algorithm is called "greedy" because it never reconsiders a previously selected city. It just goes straight ahead, traveling miles and miles without ever thinking about it.

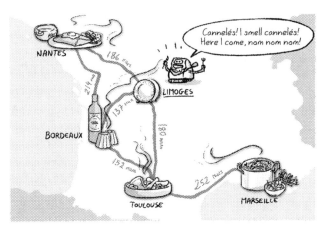

AT LEAST, IT'S FAST!

Greedy algorithms are used in numerous problems in computer science. The version we use in the traveling salesman problem is known as the *nearest neighbor algorithm* because it chooses the nearest neighbor (from among those who haven't been visited yet).

What can we say about this algorithm?

First, it effectively calculates a solution: that's a good start. Plus, it does it very fast. The program's complexity is the number of points squared, the same as with Dijkstra's algorithm, which we used for the GPS. This means the computer can work on graphs with a thousand points before hitting any real difficulty. Not so bad, right? Especially for an AI algorithm that requires less than ten lines of computer code!

Now, with that being said, the solution probably won't be the best. In certain cases, it can even be very bad.

Imagine we leave from Limoges and have to visit Toulouse, Bordeaux, Marseille, and Nantes before we return to Limoges. With the greedy algorithm, we're going to go to Bordeaux first (137 miles), then to Toulouse (152 miles), then to Marseille (252 miles), before we go 615 miles in the opposite direction to Nantes, passing through Bordeaux and Toulouse, or through Clermont-Ferrand. If we had done a little thinking, we probably would have gone to Nantes first, which is, admittedly, a little further from Limoges than Bordeaux (186 miles), but then we would have passed through all the other cities in a loop (Nantes, Bordeaux, Toulouse, Marseille), instead of making the same trip twice. Our loop would have been 805 miles instead of almost 1,160 miles!

COME ON, WE CAN DO BETTER!

The problem with the greedy algorithm is that it doesn't attempt to do any better than the first path it determines. It's a real shame because this first solution isn't necessarily very good, and it might be easy to fix a few mistakes, like going to Nantes and back, which isn't well positioned on the route.

Fortunately, artificial intelligence researchers have better methods. To understand them, we have to explore a new space.

But first let's go back to what we've already seen: *state* space exploration. Remember: state space is the set of possible states. In a GPS problem, we walk about this space in search of a solution. In the traveling salesman problem, it's exactly the same: our greedy sales representative is going to go from point to point by visiting the nearest neighbor each time.

However, there is also another space: the *solution* space. This notion is somewhat hard to understand. Imagine you're no longer concerned about the cities you pass through (the states), but instead you're concerned about the list of cities that determines your entire path (the solution). With the traveling salesman problem, any solution is one in which you've visited all the cities, no matter the order. Bordeaux, Toulouse, Marseille, Nantes is one solution. Marseille, Toulouse, Bordeaux, Nantes is another (better). All of these solutions, together, are known as the solution space.

The greedy algorithm simply searches, rather hastily, *any* solution: it chooses one single value in the solution space.

THE SOLUTION SPACE

The difficulty that arises in the traveling salesman problem comes from the size of the solution space. If we have N cities to visit in addition to our departure city, there are N choices for the first city, $N - 1$ for the next, and so on. In total, we'll have

$$N \times (N - 1) \times (N - 2) \times \cdots \times 3 \times 2 \times 1$$

solutions to visit all of the cities. This calculation, well known to mathematicians, is referred to as a *factorial*, and we write it with an exclamation point. Thus, we have $N!$ solutions for our problem of N cities. Put another way, for a state space of size N, as in the case of the traveling salesman, the solution space is size $N!$

As soon as we start talking about complexity and problem size, however, this factorial function loses its charm. Indeed, its value goes up incredibly fast. To give you an idea, if $N = 4$, we will have 24 possible solutions; if $N = 7$, that's already 5,000. And if $N = 10$, it's over 3 million! Thus, it is inconceivable to try them all to decide which is best.

ONE GRAPH CAN CONCEAL ANOTHER!

On the other hand, with all these solutions, even if there are a lot of them, we can build a graph: all we have to do is "connect" the solutions. Computer scientists love these graphs because they can write lots of algorithms with them.

To build this solution graph, we're going to connect two solutions when they are identical except for two cities. For example:

Limoges   →*Marseille*   →*Toulouse*   →*Bordeaux*   →*Nantes*

would be a neighbor solution to

Limoges   →*Bordeaux*   →*Toulouse*   →*Marseille*   →*Nantes*

We have only inverted Bordeaux and Marseille on our route.

In our solution space with 24 solutions (four cities to visit), we're going to connect each solution to six neighbor solutions, each one corresponding to a possible inversion of two cities on the route.

Now we have a graph with some neighbors, as was the case for the state space. Thus, we can use algorithms to "explore" this solution space. Why not!

SPACE EXPLORATION

This idea was used in an algorithm proposed in 1986 by Fred Glover, a researcher at the University of Colorado. It's called the *Tabu search* method.

This funny name comes from the fact that in this algorithm we are prevented from reconsidering any solution that has already been explored: it is taboo. This algorithm is not limited to the traveling salesman problem: it can be used to explore the space solutions of any AI problem.

To resolve a problem with the Tabu search method, you start with an initial solution (for example, one obtained with the greedy algorithm) and you look at its neighbors in the solution space. For each neighbor solution, you can calculate the solution's *cost* – in other words, the total number of miles you're going to have to travel (if you are trying to solve a traveling salesman problem) with this new solution. The Tabu search algorithm then chooses the neighbor that has the lowest cost. Attention: this solution may be *not as good* as your initial solution! It is simply the best one from among all the neighbors. Regardless of whether it is better or worse, you take this solution and you put the preceding solution on the "Tabu" list.

Then the algorithm continues: you examine the new solution's neighbor solutions (unless they're already taboo) and you choose the best one. Next, you put your old solution on the Tabu list and start again, and so on.

After a while, we need to stop. If we continue on like that, we may end up trying to examine all the solutions. There are too many: the famous N! For example, you say, "OK, I'll stop after I have tried 10,000 solutions." At this point, you look at all the solutions you have in your Tabu list, from the first (the one you started with) to the last (the one you ended the exploration with), and you keep the best one.

This is the Tabu method. Simple, isn't it? And this method produces pretty good results. But if we really think about it, we can do a lot better. To do so, we just need to join the efforts of more than one.

Darwin 2.0

Understanding How to Explore Space as a Group

7

A well-known AI method for exploring the solution space consists of using *evolutionary algorithms*. This method was invented in 1965 by the German researcher Ingo Rechenberg; however, artificial intelligence researchers only took an interest in it much later, in the late 1990s.

NATURAL SELECTION

The idea of evolutionary algorithms is to take direct inspiration from the theory of natural selection, the principle asserted by Charles Darwin in 1838 in his theory of evolution. According to Darwin, species evolve based on their environment. The better adapted an individual is, the more likely it is to survive. Surviving individuals reproduce and give birth to other individuals, passing on the traits that allowed them to survive (better).

Take, for example, a herd of prehistoric giraffes. The giraffes with longer legs and necks will survive better than the others because they'll be able to eat the acacia leaves that grow in the African savanna. Thus, they'll have a greater chance of having children than the others, and because their longer legs and necks are genetic traits that are passed on during reproduction, there will be more large giraffes in the next generation. In this manner, after some time there will only be large giraffes, the ones that we know today.

Artificial intelligence uses this idea to build heuristics to explore solution spaces.

I get the feeling you're a little confused. Are we going to ask giraffes to program computers? No, not exactly. We're going to write an algorithm to find the best solution from among all the possible solutions. In this sense it's like the Tabu algorithm – but this time, we won't be satisfied with just one paltry individual strolling about in the "solution space" graph: we're going to use a whole bunch. And to do this, we're going to draw inspiration from giraffes in the savanna.

COMPUTING HERDS

Imagine that, instead of giraffes, we have a herd of solutions. What I mean is, we have a cluster of points in a solution space. And imagine that it's possible to make new solutions from two solutions chosen at random from this herd. For the moment, this may seem a bit abstract, but we'll see it more clearly soon. Assume, for the moment, that this is possible.

The principle of evolutionary algorithms, then, is the following. We choose two random solutions from our cluster (the "herd" of solutions) and we make new solutions based on these two points. And we do this over and over many times. Our total number of solutions grows large with the initial solutions and all the solutions generated from other solutions.

Now, we eliminate the weakest solutions (this is the hard law of natural selection) to thin out our cluster a little. Theoretically, the solutions of this new population should be of a higher quality than the initial cluster. A few solutions from the first cluster, the best ones, will be kept. The bad ones will be eliminated. Similarly, from among the "children solutions," only the best are kept. The bad solutions are eliminated.

Then, the evolutionary algorithm gets to work on this new cluster, reproducing and selecting, so it can continue to improve the population. After some time, we should have only the very best solutions!

Only the best solutions will survive in the jungle of
evolutionary algorithms!

YOU'VE GOT TO START SOMEWHERE

To use this type of algorithm, we have to follow three steps:

```
1. Build an initial set of solutions.

2. Choose pairs of solutions to make new solutions.

3. Eliminate the weakest solutions and repeat step 2.
```

Each step presents its own difficulties. The first among them is to make a cluster with lots of different solutions. How is this possible? It took us an entire chapter to understand how to make one single solution!

Honestly, this depends on the problem in question. With the traveling salesman problem, it is very easy to build lots of solutions: all we have to do is to randomly choose one city after another. Limoges, Marseille, Bordeaux, Toulouse, Nantes, and there you have it! Done! No need to think. An algorithm that produces a solution at random can be written as follows:

```
Let L be the cluster of cities.

Randomly choose a city in L.

Remove this city from L.

Repeat.
```

It's very fast: for a graph with a hundred cities ($N = 100$), we can produce a thousand solutions in under a millisecond on one of today's computers!

Of course, there is no guarantee the randomly chosen solutions will be of good quality. Just the opposite, it's very likely they'll be totally worthless. But that doesn't matter: the algorithm will see to it that they evolve into something better.

TWO BEAUTIFUL CHILDREN!

Remember the principle: as with the giraffes, the individuals (solutions) are going to reproduce and make babies, and only the best of them will survive under the hard law of natural selection in the traveling salesmen jungle.

You're probably wondering how "solutions" that are basically just numbers in a computer can reproduce and have children. Why not have an eHarmony for solutions, while we're at it? You are right. This is just an illustration. We're going to choose solutions two by two at random and "mix" them, just like when two living beings reproduce and combine their DNA.

Let's look at an example. Imagine that we have to visit 15 cities, which we will name A, B, C, D, E, F, G, H, I, J, K, L, M, N, and O. Now, let's choose two solutions at random: Dad-solution, whose route is ABEDCFHGJMNLKIO, and Mom-solution, whose route is DFKNMJOACBEIGLO.

To build a "child" solution, we take the first cities in the order indicated by the first solution (the one we named "Dad") and the following cities in the one named "Mom."

We choose, always at random, the initial number of cities. For example, we decide to keep Dad's first six cities: ABEDCF. Next, we take Mom's cities in order (except for any we've already visited): KNMJOIGLO. This gives us the following solution:

ABEDCF KNMJOIGLO

And that's it! This operation, called a crossover, is also written as an algorithm: the parents are the clusters from which we take elements to put in the "child" cluster. If the algorithm is well written, it requires very few operations and thus allows us to develop many solutions by choosing couples of parents. If we come back to our example of one thousand solutions with one hundred cities each, by applying the cross-over algorithm to the parents chosen at random one thousand times, we get a whole new generation of one thousand "child" solutions in under a millisecond!

A LITTLE MORE RANDOMNESS

The evolutionary algorithm can also use genetic mutations, just like in real life. From time to time (at random again!), one of the children "mutates," inverting two cities. This little operation allows the algorithm to find good solutions faster, as long as these mutations remain rare. This generates solutions that are not consistent with the parents: if the result is catastrophic, the mutation will not be kept. But if it is of good quality, the mutation will benefit this piece of solutions in future generations!

An evolutionary algorithm is, thus, simply a method for exploring a solution space in a shrewd way. There is no natural selection because our "individuals" are just numbers in a machine. When you "mix" two solutions to make a new one, you're just choosing a new point in the solution space that has traits in common with the initial solutions.

If your two initial solutions are good solutions – in other words, relatively short routes for the traveling salesman problem – you have two possibilities. One possibility is that the route proposed by the "child" solution isn't as good, in which case it will be eliminated at the end of the trip. The other possibility is that the child's route is better because one of the parent's routes helped shorten the other one, in which case you keep it. This is how your population progresses.

The mutation, for its part, behaves like the Tabu search we saw at the end of the previous chapter. It allows us to explore the neighborhood in the solution space and look for new routes for our traveling salesman, without losing everything that has been calculated previously.

Researchers have mathematically shown that this type of algorithm works well. With enough trips, you get a very good solution – provided, of course, you programmed it right!

SO, WHAT IS THIS GOOD FOR?

Like the Tabu search, evolutionary algorithms are what we call *metaheuristics* because they work in the solution space, not in the state space. Like all heuristics, they don't seek the best solution: they explore just a very small part of the space as best they can to find a "good" solution within the time available. And, when you add it all up, they choose a solution that is nothing more than a heuristic: "If you pass through Nantes, Bordeaux, Toulouse, and, then, Marseille, this shouldn't be too bad."

The advantage of these methods is that they can be used for a whole bunch of different problems, no matter how the solution is built. This makes them very powerful! Metaheuristics are widely used in logistics: optimizing the routes of a fleet of trucks, managing air traffic, improving how a nuclear reactor is loaded... this is just a small sample of evolutionary algorithms and metaheuristic search methods.

There are many more examples, such as 3G mobile network optimization (i.e. where to place antennas to ensure the best possible coverage) or even the automated space antenna developed by NASA engineers, which was computed automatically using an evolutionary algorithm. The question we have yet to answer is "How do we maximize the reception?" The computer's answer is this small, twisted piece of iron wire. Who else could have invented this, if not a computer?

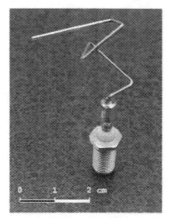

Source: Wikimedia

THIS DOESN'T SOLVE EVERYTHING, HOWEVER

Both evolutionary algorithms and the Tabu search method appear easy to use: You give the program some parameters and that's it! It figures out the solutions all by itself.

However, a lot of intelligence actually goes into choosing these parameters, for example, how they connect to neighbors in the search space for the Tabu search, how the genome is coded for individuals in evolutionary algorithms, what the right crossover method is for the solutions, and what the selection function is...

In every instance, there is always a human who tells the computer how to handle the problem. Even though the computing method is generic, the intelligence that allows the problem to be solved is entirely human.

Small but Strong

8

Understanding How Ants Find Their Way

The idea of banding together to resolve a problem isn't limited to evolutionary algorithms. Another AI method based on this principle and inspired by nature is *multi-agent systems*. Unlike evolutionary algorithms, they work directly in the state space.

MULTI-AGENT SYSTEMS

Multi-agent systems appeared in the early 1990s. They aren't just used to make artificial intelligence algorithms. At first, they were a general technique used in developing computer programs.

Imagine you want to build a city with houses, stores, transportation, and so on. You certainly aren't going to make everything all at once. You start by designing neighborhoods, then you arrange them. In computer science, we work in the same way. No one writes a program with thousands of lines of code all in one go. We develop small pieces and assemble them with each other to build a program gradually. Computer scientists work little bit by little bit.

Over time, these little bits of program have become more and more sophisticated, to the point that in the 1990s, they were no longer little bits of program but instead entire programs assembled with others. Each program in this assembly is capable of resolving a given problem but cannot function alone: it receives data from other programs and conveys the result of its calculations to another program. This small world of programs exchanges data from one end of our little computerized city to another. We call these little bits of program *agents* because they *act* on data to resolve the problem together.

A computer program written in this manner, with basic programs coupled together, is called a *multi-agent system*. This type of system is somewhat similar to ant colonies: each ant does a very specific task (find a twig, find food, move the eggs, etc.) that it repeats tirelessly for the colony to function as a whole. Each ant's work appears to be very simple and, yet, when combined together, they produce a structure that functions in a very complex manner.

THE ALGORITHM AND THE ANT

To design a multi-agent system, we have to define what the different agents do, how they exchange information and how they work together to resolve a problem. Artificial intelligence takes inspiration from this technique to write algorithms capable of resolving difficult problems, such as the traveling salesman problem.

The multi-agent algorithm used to resolve the traveling salesman problem is precisely called the "ant colony optimization" algorithm. It was invented in 1992 by the Italian researcher Marco Dorigo, who drew inspiration from the way ants move to propose a method to calculate a path.

The principle is the following: Just as with evolutionary algorithms, we're going to build a lot of solutions. Each agent is going to calculate a solution at random. But then these agents are going to exchange information about the solutions they've found:

— How many miles did you travel?
— 394.
— Oh yeah, not bad! And which cities did you pass through?
— Through Toulouse, then through Marseille. And you?
— Me, through Bordeaux first, but it was 468 miles, which isn't as good.
— Yes, this isn't as good.

Actually, the agents don't speak to each other. They just leave traces on the graph to inform the other agents of their calculations.

JUST LIKE IN THE ANT COLONY

The idea of leaving a trace on the graph is directly inspired by what ants do in nature. As ants move about, they leave behind their *pheromones*. These are small odorous molecules that help guide other ants. For an ant, pheromones work like a message: "Hey guys, look over here, I went this way." Ants like to follow other ants' pheromones. In nature, it is possible to observe columns of ants following each other: each ant is guided by the pheromones of its fellow ants.

When you place a mound of sugar near an anthill, the ants don't notice it right away. They move about at random, and one of them eventually comes across this treasure trove. It then retraces its steps back to the anthill and leaves behind a trail of pheromones to guide its fellow ants.

However, over time the pheromones evaporate. Thus, ants that take longer to return to the anthill from the mound of sugar will leave weaker pheromone trails: the odor will have enough time to evaporate before the ant gets back, and the other ants will not be so inclined to follow the same path. On the other hand, on the shortest paths, the pheromones won't have enough time to evaporate: the odor will still be strong, and it will attract even more ants. If there are several paths from the anthill to the sugar, the ants will eventually take the shortest path.

CALCULATING A PATH WITH ANTS

In computer science, the pheromones are replaced by the numbers that each bit of computer program (each *agent*) writes on the graph, directly associated with the lines connecting the points. In the computer's memory, we write a list of all the connections, and for each connection we include a cell to store a number: a cell for the direct path from Toulouse to Marseille, a cell for the direct path from Toulouse to Bordeaux, a cell for the direct path from Bordeaux to Nantes, and so on. And in each cell, the agents that pass through this connection put the total distance they traveled.

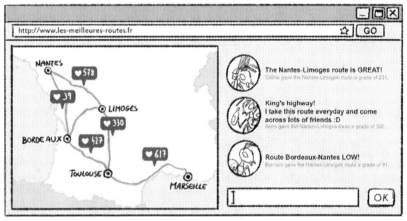

The ant algorithm

Let's come back to the dialog between the two agents: the one who traveled 394 miles and passed through Toulouse and then Marseille, and his colleague who passed through Bordeaux. The first one is going to record 394 in the Limoges-Toulouse cell, the Toulouse-Marseille cell, and all the other cells corresponding to this path. The second one is going to record 468 in the Limoges-Bordeaux cell and in all the other cells corresponding to the path that he took.

Taking a path at random like our agents doesn't take much time for a computer. Thus, we can ask a hundred agents to do this work. Each agent is going to record its total travel time on all the graph's connections for all the paths it took. Once all of the agents have finished, for each connection the algorithm calculates the *average* distance in miles traveled by the agents who passed through it.

Let's use a concrete example. Let's assume three agents – Anne, Bernard, and Celine – traveled from Marseille to Toulouse. In total, Anne traveled 260 miles;

Bernard, 275; and Celine, 210. We will, thus, put 400 (the average: 260 + 275 + 210)/3 = 248) in the Marseille-Toulouse cell.

The algorithm doesn't stop there. On the next trip, the agents go out on the road again, but this time, instead of choosing their route completely at random like the first time, they're going to take notice of the pheromones, like the ants! When they have to choose the next point in their route, they look at the values shown on the connections between the current city and the other cities. They choose the next destination at random by giving a higher probability to the cities that have connections with the lowest values. For example, an agent who is in Marseille and still has to visit Toulouse and Nantes is going to look at the weights for the Marseille-Toulouse connection and Marseille-Nantes. If the weight for Marseille-Toulouse is smaller than Marseille-Nantes, it's more likely the agent will choose Toulouse.

THE MORE THE MERRIER!

It's easy to make these random selections on a computer. All we have to do is compare numbers, do some addition, and find some averages. All this goes really fast, and our agents are going to travel across the graph in no time at all. Then we do it again: each agent records its distance in the cells, the program calculates the averages, and then everyone does it again.

For the algorithm to work well, as with the ants, the pheromones need to evaporate a little, but not right away. For this, the program stores the previous trip's average in the cell, along with the values provided by the agents for this trip. This way, if all the ants have gone every which way during a trip, there will still be some traces left on the right path.

After a few trips, the connections corresponding to the best paths will have better (pheromone) values than the others, and the agents will almost exclusively take these paths, except a few adventurous agents whose random selections lead them down the wrong path. This is just like a column of ants. There are always one or two who take a long detour to make sure a piece of sugar hasn't accidentally fallen off to the side somewhere.

THE SOLUTION EMERGES

In a multi-agent system, the solution gets built even though the algorithm does not explicitly describe how it's supposed to be built: no one has told the agents they had to find the shortest path. We're simply asking them to flip a coin when they move about and to write some numbers down.

For computer scientists, this phenomenon is known as the *emergence* of the solution. The programmer has given the agents some rather basic behaviors (here, choose a solution more or less at random by looking at what your buddies did on the previous trip), and a solution gradually emerges based on the calculations and interactions among the system's agents.

By contrast, with the Tabu method or the evolutionary algorithms we saw in the previous chapters, there is an evaluation and selection function that guides the program's choice. And even in the greedy algorithm, the heuristic is clearly described. Here, everything happens as if there were no heuristic at all and the agents found the solution all by themselves.

Obviously, that isn't the case at all. The heuristic is well hidden, but it's there: the agents have to inform each other about the length of their respective routes. It is, indeed, the humans who figured out which information the bits of program needed to exchange to build the solution.

THE SOLUTION TO ALL OUR PROBLEMS?

Just like evolutionary algorithms, greedy algorithms, or graph paths, multi-agent systems have numerous applications. The main advantage of this method is that there is no need to describe all connections between the different parts of the algorithm. The solution can be developed little bit by little bit, with each agent only focusing on its part of the problem. This method is called *distributed artificial intelligence* because the task of calculating the solution is spread out among many different agents.

Today, we use multi-agent systems in factories to control the inflow of raw materials in production chains; in telecommunications systems to coordinate network communications operations without using a central controller; and even in electric network management systems to balance supply and demand. These have become much more widespread with the advent of alternative energies (for example, wind and solar) that are not "stable" and require the network to adapt continuously. Agents that resolve the problem locally and then coordinate with other network nodes are particularly well suited for this task.

However, agents are not well suited for every type of problem: this AI technique works well only for certain types of problems in which the bits of program can make their own separate calculations, as with the traveling salesman problem.

There is no AI technique to do the cooking, wash the dishes, and go shopping. To solve each problem, we have to choose the right technique… and also provide the right heuristics!

A Bit of Tidying Up

9

Understanding How a Machine Learns to Classify

Evolutionary algorithms and multi-agent systems are two examples of artificial intelligence algorithms for which computer scientists drew inspiration from natural phenomena. There are others. Neural networks have been a big success since 2010, and they are also inspired by nature. However, before we learn how they work, we first need to get through another chapter.

FIND THE ODD ONE OUT!

Among the skills that make use of human intelligence, there is one that machines and data processing cannot do without. It is the ability to classify objects – that is, to group or separate elements based on their features.

Human beings do not have much difficulty distinguishing between birds and mammals. We learn early on how to group things that look alike: ducks, geese, and waterfowl all look like each other more than like a cat or a lion. This skill allows us to describe the personality of our friends (shy, exuberant, etc.) without needing to resort to complex reasoning.

However, classification is not as easy as it might seem.

Let's consider the five animals below: A dog, a cat, a duck, a cow, and a lion.

If you had to divide them into two clusters of similar animals, how would you group them? Naturally the cat and the lion... perhaps with the dog? And the cow and the duck together? Or would you leave the duck out, since it's not a mammal? But you could also think of the lion as the odd one out: it's the only wild animal. Or perhaps the cow: it's the only animal humans get milk from.

As you can see, grouping objects based on their similarities isn't always easy.

FROM ANIMALS TO GENES

Since the 1980s, computer scientists have taken an interest in algorithms capable of doing this same type of categorization. As an example of a concrete application of this research, biologists use computers to analyze DNA sequences and find genes shared by different species. This allows them to understand how several different DNA sequences contribute to the way a certain genetic character is expressed in a given species.

Each DNA sequence contains billions of A, C, G, and T elements. It is impossible to analyze them by hand. To accomplish this work, researchers use machines capable of grouping similar genetic data. This problem isn't that much different from the animals, except the amount of data is enormous.

This type of algorithm has numerous other applications. In medical image analysis, it is used to distinguish between areas of different tissues and to spot cancers and lesions. On video recordings, it is used to isolate an object or individual under surveillance. On e-commerce sites, it is used to make customer recommendations based on products "similar to" the ones you have already ordered.

LET'S DO SOME SORTING

Grouping and separating objects based on shared characteristics, whether animals, gene sequences, images, or books, is done using particular algorithms.

The general idea is to give the computer a large amount of data describing the objects' characteristics and to ask it to classify it all like a big boy. This is referred to as "cluster analysis" or "clustering." It consists of forming clusters of similar objects. Each cluster must contain objects similar to each other and different from objects of other clusters.

The result of this operation depends on several things, starting with the number of clusters you want to make. In the clustering example below, we've separated the animals into carnivores and herbivores:

However, if we have to make three clusters now, we probably need to think differently. For example, we could also group our animals by size:

As you can see, our clusters also depend on the characteristics of the objects. For animals, besides their food type or size, we could also consider the number of limbs, habitat, or, why not, how fast they can travel. Each characteristic could lead to a different cluster! Computer scientists call these characteristics "features," and they develop clustering algorithms that consider all the features at the same time in order to build the best cluster possible. Naturally, this requires a lot of calculations. Each object's features have to be compared to the features of all the other objects. This is precisely what computers are made for.

The algorithms that do this type of work are called *unsupervised cluster analysis algorithms*. As their name suggests, they cluster objects together based on their features, with no hint of the expected result.

IT'S ALL A MATTER OF DISTANCE

There are as many unsupervised cluster analysis algorithms as there are ways of clustering objects. For example, imagine you must divide Christmas presents among several children. You might want each child to open the same number of presents. In this case, your clustering algorithm must ensure that each cluster contains the exact same number of presents. On the other hand, if you want each child to receive presents of a similar value, regardless of how many, you'll choose a different algorithm to ensure the presents are distributed fairly. Or, if you want each child to receive a big present, you'll apply yet another clustering algorithm.

Each clustering algorithm does not sort the presents: they compare the data to cluster what they have in common. Each one has a different way of clustering a specific object's data. These algorithms have strange names: k-means, DBSCAN, expectation maximization, SLINK, and others. Each one has its own way of forming clusters.

All these algorithms compare the objects' features to form clusters of objects that are similar to each other but different from other clusters. Naturally, these characteristics (for example, the animal's size, what it eats) are represented by numbers in the computer. These are the *features*.

To describe these features, we need a computer scientist whose role is to define, for each feature, how to represent them numerically. While it's pretty easy to do this for size, choosing a number for features such as diet, habitat, or coat raises more questions. Indeed, the computer scientist must determine the distance among the different values to allow the algorithm to compare the objects. For example, is an animal with feathers "closer" to an animal with hair or an animal with scales?

Using a number to represent a measurement of distance between feature values is the basis for a great number of clustering algorithms.

In this kind of representation, each feature corresponds to a dimension in a multidimensional space, and the objects to be clustered constitute just as many points in this space. It's as if you placed your animals in the space based on their features so you could measure the distance separating them. But you have to imagine that this space doesn't just have two or three dimensions like the image above, but probably hundreds. There are as many dimensions as there are features to study!

This mathematical modeling is difficult to grasp. In any case, bear in mind that the points' relative position in the space determines the distances between the objects to be clustered. This is what will guide the clustering algorithm and determine how it regroups them. By representing the clustering problem as a geometry one, computer scientists have been able to build the majority of today's modern clustering algorithms.

STARTING OVER AGAIN AND AGAIN

The *k*-means algorithm, invented by Hugo Steinhaus in 1957, illustrates the use of this multidimensional space very well.

For it to work, you have to tell the machine how many clusters you want: this is the value *k*. The computer, then, chooses *k* objects from your list (it usually takes the first objects on the list, no questions asked). Then, it builds k clusters containing just one object each. This is just the beginning. We still have to cluster all the objects.

By using each object's position in the space, corresponding to the different features, the algorithm calculates each object's distance from the first k objects chosen. This algorithm then places each object in the cluster associated with the initial object that it is closest to. Ultimately, we get k clusters, each one of a different size. Each cluster contains the objects that were closest to one of the initially chosen objects.

This first attempt at clustering is most often of very bad quality. Imagine, for example, that we need to separate animals into two clusters and we unfortunately chose the cat and the lion first.

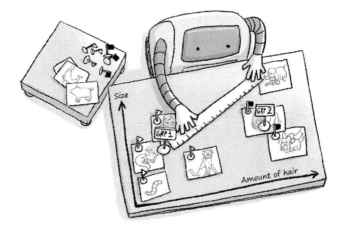

The cow, which is going to be in one of these two clusters, is farther from the cat and lion than the cat and lion are from each other!

For this reason, the k-means algorithm doesn't stop here. By using the distance function, it calculates a "mean object" for each cluster, which is the mean of all the features of all the objects in the cluster. This object doesn't actually exist (imagine the mean of a cow and a cat), but that doesn't really matter: it's simply a way of having a reference point in a multidimensional space to build a cluster. Thus, we have k "mean objects" for k clusters.

Now, we just have to start clustering all over again! The algorithm takes each object one by one, measures the distance between it and these "mean objects," and then rebuilds the clusters by associating each object with the closest mean point. With our animals, the mean point of the cluster not containing the cow is going to "catch" the missing animal (cat or lion), while the cow will remain near its mean point. In this way, we get k new clusters, which we can hope are a little better than the previous ones. For instance, the first k animals can now belong to the same cluster, which was impossible in the first round of the algorithm (remember the cat and the lion.)

However, this isn't the end of the road for our algorithm. It calculates the mean of each cluster again and hits the road again. This continues until the clusters can no longer be modified.

There is no guarantee this day will ever come, however. There is no guarantee we will get an amazing cluster either. But in practice, this works rather well. On most of the data sets, the algorithm stops after a while and produces a decent cluster. This is all we're asking of our new artificial intelligence program.

THERE IS NO RIGHT SOLUTION

There are naturally other clustering methods, such as *combinatorial optimization,* which guarantees a certain cluster quality. This isn't the case with the k-means algorithm. Like all other artificial intelligence algorithms, it's about finding a solution in a reasonable amount of time, even if it's imperfect.

The main difficulty with unsupervised cluster analysis methods is that the computer scientist must be familiar with the algorithms and with the way they cluster the data. That's because, objectively, there is no best algorithm: everything depends on what you want to do with the result. To place mailboxes in a city, for example, first you'll want to make homogeneous clusters so that each mailbox covers more or less the same surface area or the same population. But if you try to draw cities on maps, you'll use an algorithm capable of building heterogeneous clusters, because cities often have an irregular shape. Ultimately, if you are a biologist and are interested in animal species, you will most certainly use a hierarchical classification that allows you to build clusters and subclusters: cats and dogs are mammals, birds and mammals are vertebrates, and so on.

TO EACH ITS OWN METHOD

Clustering algorithms used in artificial intelligence define heuristics to build a data partition that is as consistent as possible with the user's objectives. Thus, each algorithm follows its own method: it applies its own heuristic.

Moreover, clustering algorithms work on very large quantities of data. You might be disappointed by the following information, but we do not use algorithms to sort five animals. When we use an AI algorithm, it's because the calculation is much too complicated to do it by hand, or it's because there is too much data. The problem's complexity simply makes it impossible to examine all the possible clusters: there are too many!

Because of the amount of data to be processed and because we are using heuristics, it is very rare to obtain a perfect classification. There is no way you will obtain classes of the same size even if you would like to. There will be some objects halfway between two classes that you'll have to put one place or another somewhat arbitrarily. Each time, the algorithms make choices that are not entirely satisfying. Of course, they really aren't doing the choosing: they are applying rules that allow them to come as close as possible to achieving the objectives set by the programmer.

SO, WHERE'S THE LEARNING IN ALL THIS?

Cluster analysismethods rely on the idea that the computer uses the data to build one exclusive decision rule. The result of this calculation is not limited to reorganizing a list of previously clustered objects: it subsequently allows you to assign a class to any object.

For instance, let's use the k-means algorithm, which determines the points in the feature space. If you give your system a new object to classify, the computer will immediately be able to put it in a cluster by comparing the distance to the "mean points." Thus, the algorithm hasn't just built two classes, it has also created a calculation function for classifying any object of the same family.

This gives the impression that, using the data provided, the computer has learned to classify the objects on its own. That's why computer scientists often speak of *unsupervised learning*.

Naturally, the computer hasn't learned anything: it has simply applied a classification rule built using an algorithm. This classification rule even behaves like an algorithm: we can use it to classify new objects.

JOY AND GOOD HUMOR!

It is possible to make a system "learn" plenty of things using unsupervised cluster analysis. For example, a system can learn to automatically recognize emotions on photos of faces. Indeed, by focusing on the emotional features of facial expressions, unsupervised cluster analysis algorithms are able to calculate imaginary points to represent joy or anger. As with the k-means algorithm, the machine can then use these points to recognize an emotion on a face that it has never seen before! In some way, it has "learned" to recognize emotions on its own.

However, it is important to remember that nothing happens by magic. A computer scientist has played a role at each stage of the process to define the features of the objects studied, to choose the right algorithm, and to ensure the result is used correctly. A human has told the machine which operations to run on the data provided.

The machine is not capable of finding these operations on its own because it has no imagination: it makes the calculations we ask of it, period. The computer scientist has programmed a distance function for the objects to be clustered together. The scientist has also decided how the objects will be clustered together. After that, the computer does all the work. Our computer scientist would have had quite a hard time doing this classification without the computer!

The computer surpasses humans in a good many activities, including classification. It is able to process an astronomical amount of data way faster than any human can. However, it can only do this by following the instructions provided in the algorithms.

Taking an AI by the Hand

10

Understanding That a Good Example Is Often Better Than a Long Explanation

Admit it, the idea of machine learning is kind of scary. Our natural tendency to personify machines has led us to imagine that they could be capable, like a small child, of learning from the world around them and acquiring true autonomy from their inventors, us, humans. Nothing of the sort. A machine always works within the bounds of the algorithm and the data we provide.

ASK THE PROGRAM!

To understand the principle of machine learning, we have to come back to the notion of computer programming. Computers are machines capable of processing data an algorithm described in a computer program. For its part, this program itself is provided to the computer in the form of data. Here, we can make two observations:

1. A computer uses data to produce new data.
2. A program is data.

Consequently, there is nothing to prevent the result of an algorithm – the data produced by the machine – from being a program capable of processing new data.

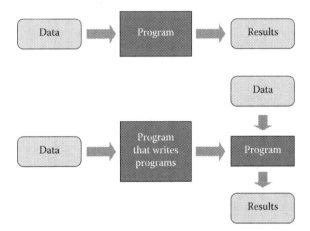

Machine learning implements this principle. This works a little as if the computer scientist gives the computer a plan (the data) and some instructions to build a new machine following this plan. The computer follows the instructions on the plan. It obtains a new machine that is also capable of processing data.

FROM DATA TO PROGRAMS

In computer science, it's a little more subtle than this. We don't exactly build new machines, but we do produce new data or new programs that allow us to process data in new ways.

Let's come back to the k-means algorithm and unsupervised classification. The computer has iterated on the data to calculate k mean points. In doing so, it has classified the data around these k points. Now if you give it a new object, a piece of data that it has never seen, it will still be able to classify it according to these k points.

Everything happens as if the algorithm has "built" a data-based classification program. This new program is capable of processing any new data of the same kind.

However, the machine does not learn on its own. For this to work, a computer scientist must first write the learning program – in other words, all of the instructions that will allow the computer to use the data to produce a new AI system. The computer cannot invent a new artificial intelligence program on its own: it just follows the procedure it receives from a human.

AN EXAMPLE IS BETTER THAN A LONG EXPLANATION

This programming technique is widely used in artificial intelligence when the problem in question is too difficult to be resolved using a heuristic written by hand. Let's take voice command systems like Apple's Siri, Google Home, or Amazon's Alexa as an example. They are able to associate phrases ("Turn the light on," "What time is it?") with specific operations: search for a definition on the internet, look up the time, turn on a light, and so on.

As we saw in chapter 2, this doesn't necessarily have anything to do with understanding what the phrase says. It's simply a matter of associating the words with the commands. One of the first systems of this type was programmed entirely "by hand" by Terry Winograd in the late 1960s. The idea was to give commands to a machine to move different-shaped objects.

This hard-to-pronounce program is called Shrdlu. To make it work, Winograd had to predict the words that would be used, the different ways of giving orders to the machine, and so on. It's long, tedious work.

The idea of machine learning is to write a program that will automatically build associations between the commands and the operation to be carried out. To achieve this operation, researchers amass giant databases, often containing hundreds of thousands of command examples. For each sentence in the database, you have to tell the machine what specific result you expect to obtain. For example:

```
what time is it  →  tell the time
```

The learning algorithm must build a system capable of associating the right answer with each possible value provided as input. In other words, we want it to "learn" to reproduce the results provided in the examples.

This technique is referred to as *supervised* learning.

THE ADVENTURE BEGINS

The first supervised learning algorithms date back to the 1950s.

At that time, Arthur Samuel was working on an artificial intelligence program using the minimax algorithm on checkers. His heuristic was based on counting the number of pieces and kings, and on the pieces' movements on the board. It occurred to him, however, that the machine could memorize the moves made and associate them with the final result (win or loss). He then wrote a new heuristic capable of considering the "experience" of previous games or previously recorded professional games when evaluating how to score each move. In this way, he used data to train his system to improve his heuristic.

Around the same time, albeit with a completely different method and objectives, Frank Rosenblatt proposed a learning algorithm capable of recognizing triangles in a picture. He gave the machine hundreds of triangle examples to calibrate a calculation function that could then determine whether a shape was a triangle or not.

These two AI programs were likely the first ones to use supervised learning algorithms.

MUCH ADO ABOUT NOTHING?

The idea of training a machine to perform a task, rather than writing how to perform the task, is not new. At first, however, it was limited to only a few applications. The machines at the time did not have enough data to use supervised learning algorithms on anything and everything.

Today, computers are capable of reacting to commands in natural language, transcribing handwriting, detecting errors in physical systems, recommending a book for you to read, and distinguishing between a face and a cat in a picture. All of this is possible thanks to the incredible amounts of labeled data that have been gathered in databases to tell the machine: "You see, in this picture, there is a cat!"

You're going to say to me: an algorithm that recognizes cats, that isn't very useful. You might be surprised to learn that kittens fooling around or looking cute for the camera are among some of the most watched videos on the internet. So, you never know how useful a computer program might be!

FROM IMAGE TO DATA

It is important to understand that the computer doesn't interpret the data. As with the Turing test, it only sees symbols, specifically, binary numbers recorded in its memory cells. As far as it can tell, these numbers could describe a picture of a cute kitten just as well as a list of telephone numbers. If you give the machine a map of Paris and you apply an animal recognition algorithm, it will recognize a cat or frog really easily!

Thus, we need to tell the learning program what the features of the data are, as we did with the unsupervised classification algorithm. This way, it's simply a matter of transforming the initial data (kitten pictures, telephone numbers, or a map of Paris) into *variables* the program can learn the rules of.

HOW ABOUT WE TAKE THE TRAIN?

The type of variables the learning algorithm must rely on is sometimes subject to intense debate. As the 1970s came to a close, the Polish researcher Ryszard Michalski wrote a learning algorithm that worked with so-called *symbolic* variables – in other words, well-formed logical formulas. The data model he used, named VL_{21}, is a *description logic*. These computer languages were very popular in the 1980s and 1990s. They let us describe variables or decision rules such as the following:

```
The animals with triangle-shaped ears are cats
```

This could be written in VL_{21} using the following formula:

```
∃ ear [shape (ear) = triangle] → cat
```

Michalski used this language to write an artificial intelligence program capable of building logic rules. This is revolutionary! The system no longer positions points in a vector space, as did *k*-means. Instead, it writes well-formed formulas that represent reasoning similar to a human's.

To illustrate this result, Michalski programmed a system capable of recognizing a series of trains. It's less amusing than kittens, but it also proves that researchers didn't lack imagination before the advent of the internet!

IT'S LOGIC!

Michalski's trains are defined by their features: number of cars, car size, whether it is open (no roof) or closed, and so on. The researcher gives his program examples of trains traveling east or west and describes their features. For example:

> *There is an eastbound train with three cars. The first car is small and has an open roof. The second car is a locomotive. The third car is small and has a closed roof.*

Obviously, all this is written in the logic language VL_{21} with ∃, [], and other abstruse symbols.

Michalski's program automatically learns, from examples, rules of the form "If there are such and such features, then the train is traveling east" (or west). In other words, if we apply the algorithm to a slightly different domain, it automatically learns, from examples, the decision-making rules that allow it to recognize objects, just like with the cat ears:

```
∃ ear[shape(ear) = triangle] → cat
```

To obtain this result, the algorithm takes the examples one after the other. At each step, it tries to build the broadest rule possible based on the preceding examples. This is possible thanks to the description logic used by Michalski. This language not only allows us to describe the variables' values (there are triangle-shaped ears) and the decision formulas (if there are triangles, then it is a cat), but also the reasoning rules for the formulas themselves. For example, when several cars have similar features, Michalski asks the machine to apply a rule that groups the examples together.

By using these logic rules, the algorithm gradually builds a decision rule that gives a correct answer for all the examples supplied to it.

TELL ME WHAT YOU READ, AND I'LL TELL YOU WHO YOU ARE

The logic rules are the essence of Michalski's algorithm. They allow us to build these final decision rules, and they have all been defined by the researcher "by hand." As always with AI, there is a certain amount of human expertise supplied to the machine. Here, the expertise has a direct impact on the learning process: if you want to learn this example, then do this or that to build your decision formula.

Still, the fact remains that the computer automatically builds decision rules in a computer language (here, the description logic VL_{21}). As with the "plan" example we used at the beginning of the chapter, the computer has built a program using examples.

All of this algorithm's power stems from the fact that the result is a set of intelligible decision rules, as long as you're speaking VL_{21}. When built in this way, the system (the program resulting from the learning) can then explain each of its decisions!

This type of approach can be particularly interesting in a recommendation system. Let's use book shopping as an example. The k-means algorithm might recommend for you a book that is "close" to other books you have chosen, thanks to a distance measurement. However, it won't be able to explain why this book is a good choice. On the other hand, a system capable of learning all the features of the books you read and grouping them together in well-formed formulas will be able to tell you: "Actually, you like 1980s science fiction books." Impressive, isn't it?

FROM SYMBOLIC TO NUMERIC

However, this supervised learning approach presents certain limitations. The biggest limitation is that, when you give the system its inputs, you need data that already have some form of symbolic representation (the car is big, the second car has an open roof, etc.) It's a little bit like when you describe an object to a blind person. You need to describe it with many details, using only words, so that the person can infer visual properties about this object without ever see them. This task gets tedious very fast, and, let's be honest, no one has fun describing kitten pictures this way.

The second problem is much deeper. As the number of examples increases, each new case leads the machine to make the final decision rule a little more complex. John McCarthy named this difficulty the *qualification problem*. Imagine you have learned that a car is a vehicle with four wheels and a steering wheel. If you encounter an example of a car with only three wheels or with a control stick instead of a steering wheel, you'll have to add these particular cases in the decision rules. For each particular case, the set of rules becomes more and more complex ... and less and less intelligible!

In the 1990s, artificial intelligence researchers therefore turned to other approaches. The machine no longer manipulates the decision rules directly. Instead, it only manipulates the variables to calibrate a program, like with the *k*-means algorithm. This type of learning has been dubbed *machine* learning.

Learning to Count

Understanding What Statistical Learning Is

Whereas the objective of symbolic learning is to build a program in a logic system (or another model), statistical machine learning only manipulates the associations between the values of variables and a result. The resulting "program" is, therefore, a set of numeric values that permits a decision to be made, like the k mean points in unsupervised classification.

But unlike the k-means, supervised learning does not give the machine the task of clustering the objects together. It defines the expected result for each input. Thus, the system does not calculate all the classes (or the answer to be given): these are described in the example databases.

In order to teach the machine to recognize cat pictures, we have to give it several pictures and tell it whether each one contains a cat or not.

NEW AND OLD ALIKE?

Statistical approaches for supervised machine learning were not invented in the 1990s: the first supervised learning models developed in the late 1950s for checkers or triangles already operated on this principle. These algorithms calculate an association between the numeric values of the inputs and the expected result.

Defenders of symbolic AI largely dominated AI research up until the 1980s. There are two reasons for this.

First, as we have seen, symbolic learning produces explainable systems because they are described in a logical language. Second, symbolic learning works well with relatively low amounts of data, and in the 1970s, we didn't have the gigantic databases that are in use today. The greater difficulty of using statistical machine learning methods was, thus, not worth it. Over the decades, advancements in computer science have helped produce much more numeric data and increased the number of variables we can work with, surpassing the limits of symbolic models.

RIDDLE ME THIS

To understand how supervised machine learning algorithms work, let's use another game as an example: "Akinator, the web genie," a software developed by two French programmers in 2007. It's a variation of the game "Guess Who," in which the computer has to guess which character the user is thinking of.

For this, the user answers questions about the character such as "Is your character a man?" The possible answers are "yes," "no," "mostly yes," "mostly no," and "I don't know." Question by question, the system ultimately identifies the person you are thinking of.

Try it out: the software is available online and on smartphones! It's rather impressive.

TREES AGAIN

To guess correctly, Akinator uses a *decision tree*, a mathematical tool that allows the choices to be programmed. The final decisions (here, the characters the player may be thinking of) are the leaves on the tree's branches. To reach them, the program follows the branches starting from the root. Each branch point represents a decision based on the values of the variables. In Akinator, these variables correspond to the questions Akinator can ask: Is the character real? Is the character a man? Is the character still living? And so forth. These questions were chosen by the game's inventors.

A decision tree is kind of like a program: you provide a set of variables describing the problem (in our example, the values corresponding to a character), and the system proceeds from branch point to branch point to find the "right" decision (in this case, the character's name).

For this to work, first we need to build a decision tree. This is a tedious task, and it's impossible to do it by hand.

A BIT OF PREDICTION

In 1986, Ross Quinlan proposed an algorithm that would go down in history: ID3. This supervised learning algorithm allows us to automatically build decision trees using examples from a database.

Each example is described by a set of attributes. ID3 then calculates the values of the branch so that the decision tree's answer is correct for all of the database's examples. The algorithm doesn't just associate the values with the result: the instructions for the branch points are calculated so as to group together as many examples as possible in the databases.

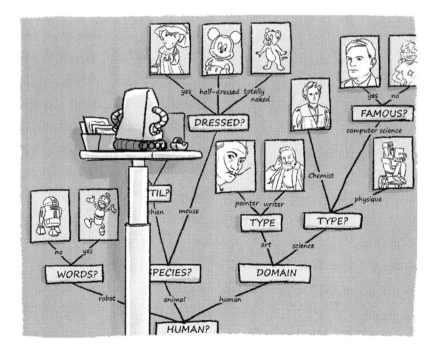

Let's assume, for example, that we have a very basic medical diagnosis database. Our database indicates the temperature, throat color, and condition (sick or healthy) of each individual. By using the ID3 algorithm, we can build a decision tree to determine whether a person is sick or in good health. The algorithm is going to determine that everyone with a temperature over 100.4 °F is sick, no matter their throat color or exact temperature value (100.8 °F, 102.37 °F, etc.). In this way, it is possible to group them together in one decision tree branch. Conversely, when the temperature is low, only individuals with a red throat are sick. This allows us to build the following decision tree:

A decision tree like this can be used to diagnose a patient who isn't in the database. Thus, if we input an individual with a temperature of 99.1 °F and a normal throat, even if there is no example with these exact values in the corpus, the algorithm will be able to confirm that the individual isn't sick!

LOTS OF EXAMPLES, LOTS OF VARIABLES

You are probably better off consulting a physician instead of relying on this prediction, because this artificial intelligence program only works with data and variables we give it. With only two variables, we can't expect a very accurate diagnosis!

The ID3 algorithm and its successors, in particular the C4.5 algorithm, were among the most widely used learning algorithms in the 2000s. Not only are they capable of building compact trees, they can also deal with exceptions when an example from a database needs to be handled in a particular manner.

To achieve this feat, at each stage of the tree, the Quinlan algorithm calculates which variable allows it to best separate all the examples it has been given. For Akinator, this means choosing the most relevant question to quickly arrive at the right answer. With the doctor, this means finding the variable and the value range that gathers the most examples. To do this, the algorithm measures the variables' *data gains* using a mathematical function. Naturally, we must calculate this value for each variable and for all the examples. However, the calculation is relatively quick, even with thousands of examples and lots of variables.

In its calculation, the algorithm also considers a few particular eventualities. If an example's attributes cannot be generalized because its values are too specific, it is excluded from the decision tree. The algorithm, then, calculates how reliable each result produced by the final decision program. Thus, it is able to tell you "with 99.1 °F and a normal throat, there is an 82% chance you are not sick." Impressive, isn't it?

TOO MUCH, IT'S TOO MUCH!

The weakness of these algorithms is the amount of data you need to build a "good" classification. The more variables you have, the more examples you need.

You're right to suspect we normally don't just use two or three variables to write an AI program. For each piece of data, there are usually several dozens of features, and sometimes many more! To interpret commands in a natural language like Siri does, for instance, we need thousands of dimensions.

Thus, the number of examples required to build a good learning algorithm can be mind-boggling. Even with the enormous databases we've built since the 1990s, the results are rarely up to the task. This algorithm is not capable of calculating general rules.

What we need, then, is another method.

THE KITTIES RETURN

The other method we need has been around since 1963. It's the support vector machine, or SVM, a model proposed by Vladimir Vapnik and Alexey Chervonenkis.

For nearly 30 years, these two Russian researchers studied how classes are built from a statistical standpoint. The two mathematicians were interested in *data distribution*. Thus, they proposed an algorithm capable of classifying data, like ID3 does, but with many fewer examples. The idea is to "separate" the data so as to have the largest margin possible on each side of a dividing "line."

To understand this principle, imagine we have a system capable of recognizing cat pictures. We'll only consider two dimensions: the number of triangles in the picture (a particular feature of cat ears) and the number of rounded shapes.

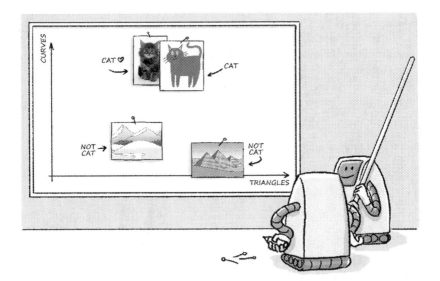

We give our algorithm the four following examples: a cat photo, a cat drawing, a mountain landscape with a hill, and a view of the pyramids.

Each picture is placed higher the more rounded shapes it has. Likewise, the more triangles it has, the further to the right it is placed.

Naturally, it is impossible to build a system capable of recognizing the pictures based on these two dimensions alone. But this simple example allows us to understand how an SVM works.

THE MATHEMATICS OF ARTIFICIAL INTELLIGENCE

The SVM's first task is to analyze the examples supplied to the system during the learning phase and draw a straight line to separate the cat pictures from the rest. In theory, this is easy. But in practice, there are many ways to separate points in a space with only one straight line, as demonstrated by our two robots in the picture.

In the first case, proposed by the robot on the left, the separating bar is very close to the photo of the cat and the photo of the pyramids, while the other two pictures are further away. By contrast, in the second case, the bar is really close to the picture of the mountain and the drawing of the cat. And between the two, there are many other possible positions. How do we choose?

The first thing we can see is that these two "extreme" bar positions each pose a problem. Indeed, our objective isn't simply to separate the pictures hanging on the board but to build a program capable of sorting through

new cat and noncat pictures, without knowing which category they belong to ahead of time. If we choose the first dividing line, the one that comes near the photos, we have a high risk of error. Indeed, a very similar cat photo having slightly fewer curves or triangles will go on the other side of the line and end up being classified in the noncat category. By contrast, a picture of pyramids with slightly more curves would be considered a cat.

Similarly, the solution proposed by the other robot, which comes very close to the cat drawing, is going to produce an algorithm that also makes mistakes easily. In the end, Vapnik proposed choosing the solution that provided the greatest margin possible from one side of the dividing line to the other, such that the two groups would be as far as possible from the line. Using mathematical calculations, the algorithm thus finds the straight line with the greatest margin. This increases its chances of correctly classifying future pictures.

This algorithm is called the optimal margin separator. In the drawing with the third robot holding the bar, this means having the largest possible gray strip from one side of the separator to the other.

TOO MUCH CLASS!

Using calculations, it is thus possible to separate the dimension space into two parts: the "cats" side and the "noncats" side, in our example.

And the benefit is, once the algorithm has used the examples to separate the two sides, it can define a decision system. The system is then able to classify any described object based on its attributes into the "cat" category or the "noncat" category, even if it is not included among the learning examples.

This calculation even works with very little data: the algorithm systematically builds the optimal margin, regardless of how many dimensions and examples are provided. Naturally, the more examples there are, the more reliable the separator will be.

This learning algorithm is, therefore, particularly interesting. Even if it requires complex calculations to find the separator, it allows us to handle problems with many dimensions and relatively little data. By contrast, this is not the case with algorithms that use decision trees.

KEEP STRAIGHT

However, even in the 1990s, no one really gave much thought to this method.

The SVM algorithm does, indeed, have a serious limitation: the separation that it establishes between the two clusters is inevitably *linear*, in the mathematical sense of the term. In plain language, it's a straight line.

Let us consider the picture below:

It has a lot of curves and triangles, which means it would be placed in the top right of our dimension space – that is, next to the cats. And there, it's no good: to separate this picture from the cat group, we would need a curved line – in other words, a *nonlinear* separator. However, the SVM algorithm is unable to produce this kind of separator. It's rather frustrating!

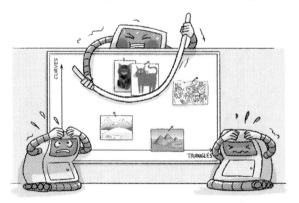

THE SVM STRIKES BACK

Nevertheless, in 1992, Vladimir Vapnik, Bernhard Boser, and Isabelle Guyon wrote an article that would revolutionize machine learning. Using a mathematical transformation called the "kernel trick," they showed that it was possible to produce nonlinear classifications with an SVM.

The *kernel* is a function that measures the similarities of two points in a feature space. This is analogous to the distance function used in the *k*-means algorithm. With the SVM, the *kernel* allows us to calculate the margin and, thus, the separator.

Vapnik and his colleagues had a great idea: add imaginary features to the kernel to force the algorithm to work in a space with more dimensions. It's kind of like some physicists were to say to us: "OK, ultimately, the third dimension isn't enough. Based on our calculations, we going to pretend the space has eight dimensions and we'll see what we get as a result."

This *kernel trick* allows us to break free of the SVM's limitations: the algorithm still calculates a linear dividing line, but since it is formed in a space with greater dimensions, it is no longer linear when we come back to our initial dimensions.

It's a little hard to imagine, isn't it? Think of our physicists. For years, they drew triangles using straight lines. Then, one day, they realized that Earth was round and, as a result, all the triangles were actually distorted when they would draw Earth on a piece of paper. This is more or less the idea behind the *kernel trick*.

INTELLIGENT, DID YOU SAY INTELLIGENT?

The SVM has had a lot of success in computing since the dawn of the 21st century. It is very effective at recognizing cat pictures, even when there are dirty tricks like a wolf with its ears up or the Pyramid of Giza (don't laugh, the triangular shape tricks the computer very easily). Let's be serious; SVM is particularly very useful for a whole ton of other applications, starting with understanding natural language (human language).

Processing natural language is a difficult problem for AI and, yet, one of the most ambitious problems to solve. Indeed, language is a very complex phenomenon that is impossible to describe using symbolic rules alone. As soon as you leave a very specific framework, as soon as the vocabulary becomes the slightest bit realistic, the rules become too abundant.

The kernel tricks consists of adding a dimension.

It is not possible to use decision trees either: the classification algorithms need a lot of examples to cover each dimension. However, sometimes there can be just as many dimensions as words! In the tens of thousands! The SVM allows us to circumvent this limitation and build classifications for objects with many variables and many values. This isn't some kind of miracle, though. For this to work, we need data. A lot of data. Positive examples (this picture contains a cat). And negative examples (this picture does not contain a cat).

If Google's applications are currently the best at document searches, image recognition, and machine translation, it's simply because they have colossal amounts of data that their competitors do not! Try to imagine how much time and energy it takes to build a database with a label for each piece of data: this is a cat, this is not a cat. Thanks to the billions of searches we do every minute on their servers, Google researchers obtain this data for free. You have to admit that it's worth it!

CAREFUL, DON'T SWALLOW THE KERNEL!

However, just like every AI technique, the SVM has a weakness. To work well, an SVM not only needs the right data set, it also needs a kernel function that is well suited to the problem to be solved. This means that a computer scientist must not only define the problem's dimensions (the features) but also what happens inside the kernel. This is a bit like the heuristic in the minimax method for chess: if you don't tell the machine how to evaluate the board, it won't play correctly.

There are some classic functions for the kernel that are well suited for each kind of problem. They are easy to control and generally produce good results. But as soon as you try to increase the AI's performance, it's not enough to just give it more examples: you have to look under the algorithm's hood. In other words, you have to hot-rod the kernel.

Many years of higher learning in mathematics and computer science are needed to understand how an SVM works and to be able to manipulate these algorithms. Researchers spend several years building one kernel for a specific problem. However, advancements in AI image processing and natural language understanding, two major artificial intelligence tasks, have been lightning fast since the advent of the SVM.

In the early 2010s, however, there was a new revolution in machine learning: artificial neural networks.

Learning to Read

12

Understanding What a Neural Network Is

The idea of building artificial neurons is not new. It dates back to the middle of the 20th century and understanding how the human brain works. In 1949, Donald Hebb, a Canadian psychologist, hypothesized that brain cells called neurons are the basis of the brain's learning mechanism. He made a fundamental assertion: if two neurons are active at the same time, the synapses between them are strengthened.

Today, neurobiologists know that it isn't that simple, but the "Hebb rule" has paved the way for important neuroscience research and served as a foundation for artificial neural network development.

DRAW ME A NEURON

An artificial neuron is nothing like a human neuron. It's simply a term used to refer to a number-based computing operation. The computed result depends on several neuron input *parameters*.

Let's take a closer look.

To begin with, the "neuron's" input consists of a list of numbers, which are usually binary (0 or 1). Each number is multiplied by a coefficient, called a *weight*, and the first step of the calculation is to sum everything up. Mathematicians refer to this as the *weighted sum*.

The second step is to compare this sum with a value, called a *bias*. If the sum is greater, the neuron's output is a value of 1. If this is the case, we say that the neuron fires up. If not, its output is 0. Note, however, that there is no object or light. This is simply a calculation that produces a certain result.

The weight and bias are the neuron's parameters. They allow us to control how it works.

For example, let's take the "neuron" below. We've indicated the inputs, the weight and the bias.

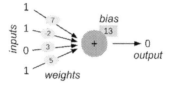

In this example, the sum calculated by the neuron is:

$$1 \times 7 + 1 \times (-2) + 0 \times 3 + 1 \times 5 = 10$$

The result (10) is smaller than the bias (13). Thus, the neuron's output is 0 (as a result, we say that it remains "off"). This is the result of the calculation.

By way of comparison, here's a diagram of a human neuron from an encyclopedia. You have to admit, this is totally different!

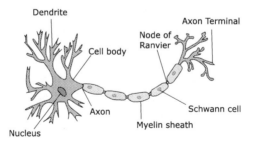

MORE TRIANGLES

An artificial neural network is an algorithm that makes this type of "neural" calculations on data.

The first networks were programmed in the mid-1950s to simulate human neurons, as in Hebb's theory. This is how "neural" came to be used to describe this type of calculation. However, artificial neural networks really get their start with the *perceptron* algorithm, which was invented by Frank Rosenblatt in 1957.

By seizing on the idea of artificial neurons, Rosenblatt programmed an algorithm capable of learning to recognize black-and-white triangle pictures. What made the perceptron unique was that the calculations were organized in consecutive "layers" of neurons, as in the drawing below:

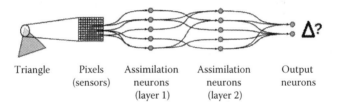

| Triangle | Pixels (sensors) | Assimilation neurons (layer 1) | Assimilation neurons (layer 2) | Output neurons |

To the far left of the diagram, sensors transform the picture into numbers: each *pixel*[1] is transformed into a 0 if it is off or a 1 if it is on (in gray in our drawing). The machine used by Rosenblatt for his experiments had a resolution of 400 pixels (20 × 20), which is ridiculously small compared to today's photographs. The perceptron's input is thus 400 numbers (0 or 1), which correspond to the picture's pixels.

The perceptron's first layer is composed of several hundred neurons that each calculate a different result using these numbers. Each neuron has its own weight and its own bias, which makes all of the results different. From the first "layer" of calculations, we obtain several hundred numbers, which are either 0 or 1. For Rosenblatt this operation is the *assimilation*, and this first list of calculations is the "assimilation layer."

All of these results are then fed into a second input layer. Again, hundreds of different calculations produce hundreds of results, which are either

---

1   In a computer, the picture is cut up into little dots called *pixels*. To understand what a pixel is, stick your nose up against a television screen or a poster. You'll see small colored dots that, from afar, make up the picture. Telephones, televisions, computer screens, and internet images are all described by their *resolution*, which is expressed as a number of pixels. This indicates the picture quality.

0 or 1. We could continue on like this for quite some time, but Rosenblatt, in an article published in *Psychological Review* in 1958, proposed stopping after two assimilation layers.

The third and final layer contains only one neuron. It takes the numbers calculated by the preceding layer and runs a "neural" calculation to obtain a 0 or a 1. The goal of this little game is for the final output to be a 1 when the picture contains a triangle and a 0 in all other cases.

JUST A LITTLE FINE TUNING

Unfortunately, it's not that simple: the real difficulty consists of finding good values for each parameter and bias. It's like a gigantic mixing console with millions of mixing knobs that need to be turned to just the right position.

For each neuron, you have as many knobs as inputs. Imagine that the first layer of our perceptron contains 1,000 neurons; that makes 400,000 parameters to adjust, just for the first layer, and each one can be any value. There is no way we're going to just happen upon the right configuration by chance.

This is where supervised learning comes in.

A LONG LEARNING PROCESS

The learning algorithm allows us to automatically calculate the values of the parameters that leads to the right answer: 1 when the picture provided is a triangle, and 0 if not.

At the start, all the parameters are set to 0. For each example, the network calculates an output: it either fires up the "triangle" neuron or it doesn't.

If the answer is right (e.g. the triangle neuron fires up when the picture provided is indeed a triangle), we don't touch anything. On the other hand, if the calculation is wrong, we need to adjust the weight and the bias. To do this, we're going to increase or decrease by 1 all the parameters that were used to calculate the incorrect output.

Concretely, if the network answered "yes" even though we showed the machine a round object, we look at all the neurons in the network that fired. Then, we lower by 1 all the weights that had an input of 1. But we don't touch the ones that had 0, since they didn't affect the neuron's output.

Let's come back to our example with four values:

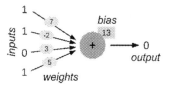

No matter what the value of the third weight is, the sum will always be 10. When the input is 0, the weight isn't used in the calculation.

Thus, we decrease the neural network's weights, but we also increase by 1 the bias of each neuron that fired so that they will be a little harder to fire up with the values provided. The idea is to make the last neuron not fire anymore if we feed it the same picture again.

Conversely, if the network answers "no" even when the operator fed it a triangle, we have to increase the weights of the nonzero inputs and decrease the bias for the neurons that didn't fire (to reduce how much they filter).

Now, we make the same adjustments (move the sliders by 1 in one direction or the other) to each example in the learning database. Obviously, we don't do this by hand. The weights are just numbers in a computer, and the algorithm does the calculations all on its own by considering thousands of previously memorized examples one after another. After a good number of examples, with a bit of luck, the knobs are in a position that allows them to recognize triangles, cats, or something else, based on what we decided to teach it.

WHERE DID MY BRAIN GO?

Rosenblatt's machine used less than a thousand "neurons"; however, that still leaves hundreds of thousands of calculations to do for each example. Despite that, this algorithm is generic enough to resolve many artificial intelligence problems, and the first results on triangles were very encouraging. Rosenblatt firmly believed that perceptrons could one day learn, make decisions, and translate texts into foreign languages.

The euphoria didn't last. In 1969, two renowned researchers published a book that brought this type of research to a screeching halt. They were none other than Marvin Minsky, one of the fathers of AI, whom we've already talked about, and Seymour Papert, another big name in computer science, to whom we're indebted for Logo, an educational programming language.

In their book, these two eminent researchers described the perceptron's advantages and disadvantages. Among other things, they demonstrated that a "neural" calculation was incapable of resolving simple problems, like the famous "nonlinear problems" mentioned in the previous chapter. As they put it, a neuron with two inputs could not be adjusted to fire only when one of the inputs alone is 1 and the other is 0: this is what we call the "exclusive or problem."

It is true: a "neural" calculation by itself is not capable of calculating a nonlinear function... but by layering the neurons as Rosenblatt proposed, even with only one assimilation layer, it is possible to circumvent this limitation and calculate just about any function. In fact, Rosenblatt's triangle recognition problem is nonlinear, and it was solved with this neural model.

Unfortunately, many researchers misunderstood Minsky and Papert's book. The "exclusive or" example was cited indiscriminately in numerous scientific articles, leading many to be skeptical of the perceptron. This despite the fact that, by 1972, several publications in scientific journals demonstrated how to resolve an "exclusive or" problem with a two-layer perceptron. Unfortunately, it would be another ten years before anyone showed any interest in artificial neural networks again!

ONE MORE LAYER!

Neural networks would eventually come back in vogue in the early 1980s. As so often happens in science when a theory starts to fade into memory, it was ultimately reinvented. The physicist John Joseph Hopfield rediscovered neural networks in 1982. He proposed a new model that, in the end, was not so different from Rosenblatt's perceptron and breathed new life into this scientific topic.

In 1986, the psychologist David Rumelhart and the computer scientists Geoffrey Hinton and Ronald Williams proposed the notion of *back-propagation*, a new learning algorithm inspired by the algorithms of the 1960s, to adjust the parameters of mechanical controllers. Using some calculations, the algorithm determines which parameters have contributed the most to the network's right, or wrong, answer. It then adjusts the weights and biases of these parameters more aggressively, and adjusts less aggressively for the parameters that contributed "little" to the result.

In this way, the algorithm finds the right parameter values much more quickly.

NETWORK SUCCESS

Thanks to back-propagation, artificial neural networks made phenomenal progress! They were quickly adopted in various domains to automate classification tasks, to control robots, to predict market changes, to simulate how hydraulic systems operate, and more.

The application we most often see in our daily life might very well be automatic character recognition, which allows banks to verify check amounts or the post office to sort mail by zip code. Most often, the machines used are programmed with a single-layer perceptron and a back-propagation algorithm.

Up to the mid-1990s, neural networks were all the rage. However, they remained difficult to manipulate. It takes years of training to be able to configure a neural network for a given task. Researchers often learn by trial and error and must use all of their expertise to determine the number of neurons an input layer requires. You need to have just enough to calculate the nonlinear function, but not too many!

Indeed, each additional neuron increases the number of calculations the machine must make. If you build an input layer from one thousand neurons that each receive a thousand inputs, that's already one million additions and just as many multiplications. One iteration of an algorithm only takes one second... but you have to repeat it on millions of examples to determine the weight.

ONE MORE SMALL BREAK

Neural networks, as we can see, require lots of calculations to be trained correctly. This is one reason why Vapnik's SVM (the support vector machine we saw in the previous chapter) replaced neural networks in the 1990s. Because SVMs are much simpler to manipulate, they obtain better results for classification tasks and, rather quickly, for most supervised learning problems. By comparison, neural networks are limited by a computer's processing power.

Nevertheless, this would all change in the late 2000s with the advent of a new calculation tool.

Learning as You Draw

13

Understanding What Deep Learning Is

In the mid-2010s, a new type of computation appeared: graphics processing unit (GPU) computation. This was a genuine revolution in the world of machine learning.

ARE VIDEO GAMES BAD FOR YOUR HEALTH?

A computer is made of numerous electronic components: the processor (or CPU) that does the computation, the memory that stores the data, but also the keyboard, screen, mouse, hard disk, and so on. All these components work with binary numbers computation. Thus, the image you see on your screen is the result of operations carried out by your computer to determine which pixels should be lit up and which should not.

When you read a web page or work on your tablet, the display requires few operations: all of the computations can be done by the computer's processor. But when you play a video game, the computer has to continually process new images to display them, and it has to delegate this work to a specific component: the graphics card.

Graphics cards are an essential component in video game consoles and, just as often, computers. Each new generation offers better image quality for a better gaming experience: it's the graphics cards that make this possible. The performance of these components has continuously increased since the 1980s, driven by the formidable economic clout of the video game market.

PARALLEL COMPUTING

A graphics card is made up of hundreds (even thousands) of small mini-processors: "graphics processing units" or GPUs. The processors are considerably less powerful than the computer's main processor, but they work "in parallel" – that is, all at the same time.

Imagine you're playing a tennis match. The ball is a yellow sphere that must be displayed on the screen 60 times per second. To determine which pixels to color yellow on the screen, the graphics card has to compute every 16 milliseconds the ball's position and its relative size. This is done by carrying out relatively fast calculations on mathematical objects: matrices and vectors. Simply put, we can say that this is just some additions and multiplications carried out repeatedly on numbers stored in the computer's memory.

The thousands of processing units carry out all these operations at the same time for the image's million pixels, which allows the display to be updated very quickly. This is only possible because all the small calculations on the pixels can be carried out independently from one another. You don't need to know the first pixel's result to calculate its neighbor's. Computer scientists call this *parallel processing.*

THE RETURN OF THE NEURAL NETWORKS

In the early 2010s, AI researchers started to take an interest in graphics cards, which were initially designed for video games and films. For a few hundred euros, it is possible to acquire thousands of parallel computing units. True, these mini-processors run very simple calculations, but they do them all at the same time, which is very fast!

Some AI algorithms, such as the Tabu algorithm we used for the traveling salesman problem, are unable to make much use of parallel processing. Each stage of the algorithm uses the result from the preceding stage: the calculations are not independent. By contrast, other algorithms, such as neural networks, can be run in parallel. The calculation of a layer's first neuron does not need the calculation from the layer's second neuron: the two calculations are independent.

Plus, the calculations carried out by the neural network are simple: additions and multiplications. The GPU's small processing units are perfect for the job! And that changes everything.

When you work with only one processor, you cannot build a neural network with more than one layer of a thousand neurons. However, thanks to the little GPUs working in parallel, it is possible to design networks with several layers ... and practically as many neurons as you want on each layer!

Deep learning is the name researchers have given to these neural network algorithms with multiple assimilation layers, as opposed to the perceptron, which is limited to only one assimilation layer.

WHY ADD MORE?

It isn't necessary to descend into the depths of the abyss to resolve AI problems using neural networks. With just two assimilation layers, you already have a deep network.

In theory, adding additional layers does not allow you to calculate new things. Mathematicians have shown that a network with only one assimilation layer could learn the same thing as a deep network ... provided it has enough neurons. However, in practice it's not that simple.

Building a "deep neural network" algorithm accelerates convergence in the learning process; in other words, it boosts the speed with which you obtain good positions for the network's adjustable knobs. Recall that our goal is to adjust thousands of parameters in order to correctly respond to an input. This is done by progressively adjusting the parameters'' values based on the examples in the learning phase.

In a deep network, the "right" values for each layer's parameters depend on the preceding layer's work. Everything happens as if you have broken down your big problem into a series of calculations that depend on each other but are simpler to adjust. By breaking the problem down, and combining it with a good back-propagation algorithm, it takes much less time to determine weight values that will produce good results.

THE CRUX OF THE PROBLEM

Deep neural networks are a win-win. Not only are they no longer limited by the computer's processing capacity (thanks to the GPUs, which allow it to run operations in parallel), but their layered arrangement allows us to obtain better learning results. What happiness!

Of course, we have to determine how to configure the network – in other words, the way in which the calculations are performed. For this, researchers build several configurations with two, three, four, ten, or sixteen different-sized layers. Then they compare the performance of these different networks on the learning database so they can choose a configuration that produces the best results. This very empirical step is done automatically. The computer compares the outputs of the different configurations, based on the data provided by the human, by using the exploration algorithms written by the human.

So, what's the problem? Why isn't everything resolved by these machine-built neurons?

The machine learning system's performance isn't so much a result of the chosen configuration as it is a result of the way the data are built and analyzed by your algorithm. Let's look at an example.

To allow a machine to "read" a word written by hand, you can use a neural network to decode each character one by one. However, your system will certainly be more efficient if it can also consider neighboring characters, since the symbols aren't completely independent. Thus, you want to build a neural network whose inputs are a combination of letters, as opposed to a single letter.

This combination is done using mathematical calculations that also use parameters. Modern algorithms use machine learning to adjust these parameters. In this way, you have neurons that not only learn the parameters to answer the question correctly but also learn the *hyperparameters* to combine data or to determine the network's configuration.

All this work performed by the machine must first be thought out by the computer scientists: the humans determine how to combine the data to obtain better results based on whether the task to be resolved needs to understand words, recognize a picture, avoid an obstacle, and so on. Each problem requires us to build the right solution, and good results are hard to reproduce on another problem, even if it is similar.

THE ACHIEVEMENTS OF DEEP LEARNING

Thanks to deep learning, artificial intelligence has made mind-boggling progress in very difficult areas such as image analysis and natural language processing. It is also used in AlphaGo, the Go program written by Google capable of beating any human player. Most surprising of all is the speed with which big-name companies like Google, Facebook, and Apple, just to name a few, have taken to the subject.

When a new algorithm is invented by researchers, it often takes decades before businesses can use it in their systems. For deep neural networks, each new invention leads to another new invention in less than a year! This allows machine learning to grow at lightning speed. And obtain impressive results! ... and hit limitations!

The primary limitation of neural networks is their inability to explain the decision they've made. This problem was raised at the start of machine learning by artificial intelligence researchers. It is still true.

The machine makes calculations on the input to produce a correct output in 99% of cases. However, it cannot explain why it has produced a correct answer, or even how. No one is able to explain what its decision is based on, which is bothersome if we want to understand and control what happens in the machine.

"Symbolic" algorithms like Michalski's, which we saw in chapter 10, seek to provide a "logical" explanation for each decision the system makes. By contrast, a perceptron is unable to affirm that a triangle is a three-sided shape! It recognizes the triangles, that's all. Similarly, a self-driving car brakes to make a turn but it cannot explain "why."

WATCH OUT FOR TRICKS!

Another difficulty with neural networks is that it is hard to solve problems if there are no learning examples. Researchers hypothesize that a well-trained network works more or less "continuously": an input similar to a correctly processed example will also be processed correctly by the network. Unfortunately, that isn't the case.

Many research teams have had a fun time inventing algorithms capable of building examples that cause neural networks to fail by making very small changes to the initial inputs. They call these "adversarial patches." These patches can be inserted in an image to make any neural network algorithm fail. At the top of the next page is an example from an article written by Google researchers in 2018. The system recognizes a toaster instead of a banana when the adversarial patch (the little round picture) is added.

TO EACH HIS OWN METHOD

The support vector machines and decision trees we saw in the preceding chapters are not susceptible to these kinds of tricks. On the other hand, they require sophisticated input transformation algorithms to process the inputs' features. This work is difficult and can only be done by a human who has spent years studying AI.

An example of a neural network "trick"
Source: arXiv.org

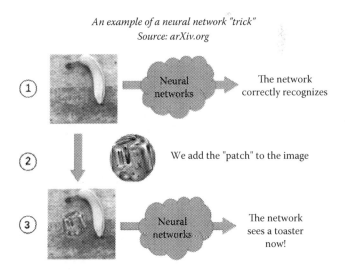

In neural networks, feature processing is incorporated into the learning process, which makes the network easier to use ... but also susceptible to error.

No AI algorithm is perfect. Each AI method has its advantages and disadvantages. The secret is knowing how to use them right.

Winning at Go 14

Understanding What Reinforcement Learning Is

Go is kind of like the "Grail" for artificial intelligence researchers. Most of the "combinatorial games" we saw in chapter 5, like chess or checkers, are resolved by a computer using the minimax algorithm. Not Go, though.

THE QUEST FOR THE GRAIL

A Go board has 361 intersections where you can place a white or a black stone. That makes 3^{361} possible game situations. Written out, that number is:

    17408965065903192790718823807056436794660272495
    02635411948281187068010516761846498411627928898
    87149386120969888163207806137549871813550931295
    14803369660572893075468180597603

To give you an idea of the magnitude, let's consider the number of atoms in the universe (10^{80}), which is pretty much the gold standard for big numbers in artificial intelligence. Now, imagine that our universe is but one atom in a larger universe. You make a universe containing as many universes as our universe contains atoms. And you count the number of atoms in this universe of universes. That adds up to lots and lots and lots of atoms. It's enough to make you dizzy.

To have as many atoms as there are situations in Go, you need to make a trillion universes like this! In short, we are very far from any modern computer's processing capacity. As with chess, we have to use an AI algorithm.

Unfortunately, the minimax algorithm isn't going to do it for us.

A BIT OF CALCULATION

Remember, this algorithm relies on the idea of trying all the game possibilities on a few moves, and choosing the one that leads to the best result. With 361 possible moves at the start of the game, the number of situations to examine increases very quickly. For instance, to evaluate all four-move games alone, the minimax algorithm must consider $361 \times 360 \times 359 \times 358$ possibilities – more than 16 billion configurations. It will take several seconds to process this with a modern computer.

With only four moves examined, the minimax algorithm will play like a novice. A game of Go consists of several hundred moves. Given the billions of possible situations after only four moves, it is not possible to see your opponent's strategy at all! We need to come up with a better method.

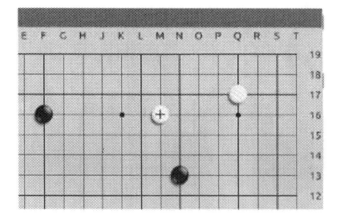

In Go, players take turns placing black and white stones on the board to form territories. The white player just made his second move. It is hard to say exactly how the game will turn out at this point.

WANT TO PLAY AGAIN?

The algorithm used by computers to win at Go has been known since the late 1940s by the name "Monte Carlo Tree Search," a reference to the casino located in Monaco.

What an odd name for an algorithm! It was given this name because this algorithm relies on chance to resolve problems.

For each of the 361 possible moves, we're going to play a thousand games at random. Completely at random. The opponent's moves and the computer's responses are drawn at random. We make note of the game's outcome, and we start over one thousand times. Each time, we start over with the same initial configuration.

To evaluate each of the 361 moves, the algorithm makes a statistical calculation on the scores obtained for the thousand games played at random. For example, you can calculate the mean of the scores (in practice, computer scientists use slightly more complex calculations). The Monte Carlo algorithm makes the hypothesis that, with some luck, the best move to make is the one that leads the board to produce the best statistics. This is the move that will be chosen.

By assuming that each game lasts under 300 moves (which is a reasonable limit if we base this on games played by humans), we will have made a maximum of 300,000 calculations for each of the 361 "first moves" that we have to evaluate, which is around 130 million operations in total. This is more than reasonable for a modern computer.

RED, ODD, AND LOW!

Monte Carlo algorithms have been used in this way since the 1990s to win at Go. In 2008, the program MoGo, written by researchers at Inria, the French Institute for Research in Digital Science and Technology, achieved a new feat by repeatedly beating professional players. By 2015, programs had already made a lot of progress in official competitions, although they had failed to beat the champions.

However, in March 2016, the program AlphaGo defeated the world champion, Lee Sedol, four matches to one. For the first time, a computer achieved the ninth Dan, the highest level in Go, and was ranked among the best players in the world!

To achieve this feat, the engineers at Google used a Monte Carlo algorithm combined with deep neural networks.

As with any Monte Carlo algorithm, lots of matches are played at random to determine the best move. However, the computer scientists do not make the computer play completely at random. Not only is the computer unable to truly choose at random (there is always some calculation), but it would also be unfortunate to play completely at random, since even the greenest Go players know how to spot the worst moves.

This is where AlphaGo's first neural network comes in. The Google team refers to this as the "policy network." This is about teaching a neural network how to detect the worst moves by giving it game examples as input. This prevents the Monte Carlo algorithm from producing too many bad matches and improves the accuracy of its statistical calculations.

A deep neural network doesn't work much differently than Rosenblatt's perceptron: the input is a 19 × 19 three-color grid (black, white, and "empty"), and the output, instead of a simple "yes" or "no," is a value for each possible move. With a deep neural network and some graphics cards to do some GPU processing, this is feasible.

NO NEED TO BE STUBBORN

The other improvement AlphaGo makes consists of stopping before the game is over, since often there is no point in continuing when it becomes clear the game is going so poorly for one of the players. Once again, this is a problem that is well suited for a neural network, where the input is the state of the board and the output is a yes or no answer to the question "Is the game completely lost for one of the players?" The Google engineers call this the "value network" – that is, a network that measures the value of the game.

The interest in ending the game before it's over is to play shorter games ... and thus play more every time. Instead of only one thousand games for each move, the algorithm can simulate two to three times more ... and the statistical calculation to determine the best move will be even better.

A BIT OF REINFORCEMENT

By combining a Monte Carlo algorithm with neural network algorithms, the AlphaGo team was able to program a computer that quickly and correctly decides what move to make.

However, to program these two neural networks, the AlphaGo team has to have examples of Go matches. Indeed, it's not enough to show the computer a Go board and tell it "Go on, learn!" The policy network has to be told which configurations contain good moves and which moves are to be avoided. The value network has to be shown thousands of Go matches and told, every time, which ones can be chalked up as a loss and which ones are sure to be a victory.

Whatever learning method you use, you have to provide the machine with lots of examples. It is not possible to find all the examples manually by searching in the Go archives. The solution used by the AlphaGo team relies on another artificial intelligence technique: *reinforcement learning*.

STRONGER THAN EVER!

It is hard to say exactly when reinforcement learning was invented. Even so, its principle plays an important role in AI algorithms. The TD-lambda algorithm invented by Richard Sutton in 1988 is widely considered the first reinforcement learning algorithm that could be broadly applied to many categories of problems.

All reinforcement learning algorithms share a common feature: they do not need examples to learn.

When young children learn to catch a ball, they learn by trial and error. They move their hands without much rhyme or reason at first, but they gradually learn the right movements to sync up with the ball's movement. In this learning process, children don't rely on examples. They correct their errors over subsequent attempts.

This is the idea behind reinforcement learning: the computer calculates a "score" for each possible action in each situation by "trial and error." As it does this, it gradually adjusts the calculation's parameters.

STEP BY STEP

Like back-propagation in a neural network, reinforcement learning lets us adjust the parameters of a sequence of calculations. However, unlike supervised learning, the output doesn't depend on examples. The systems that these algorithms work on don't have input, per se. They begin at an initial state and perform a sequence of actions to reach a desired state.

Let's use Go as an example. The initial state is the empty board. The system makes a series of moves, observes each of the opponent's responses, and the final state is achieved when the game is over. On the basis of the result at the end of the game, you can assign a score to your series of moves. Computer scientists call this a *reward function*.

In this type of system, there is no corpus of examples associated with the correct output, as is the case with neural networks. We have a given sequence of actions and just as many intermediate inputs. In this way, each state on the game board corresponds to an input, and the system's response is the action chosen by the system (its move). The actions are only evaluated at the end of the game, once all the actions are complete.

Reinforcement learning consists of adjusting the parameters of these actions by using a reward function. Each action, or each intermediate state (this depends on which algorithm you use), is given a score. When the computer reaches the final state, the algorithm adjusts the scores of all the actions or intermediate states encountered during the match in accordance with the reward value. For example, imagine that the system's first move is H5, its second is B7, and so on until the end of the match, which it wins handily. In this case, the scores for actions H5, B7, and the following moves will be increased. By contrast, if the match results in a loss, their scores will be decreased by however much the reward function decides.

To adjust the parameters of your system, which can decide which action to take at each stage of the process, the reinforcement learning algorithm is going to play millions of "fake" Go matches to adjust its parameters. To ensure that the learning is effective, this algorithm carefully chooses its actions. Unlike with the Monte Carlo algorithm, its moves are not made completely at random. As the algorithm advances, the "right" parameters have a higher chance of being chosen. In this way, the system *reinforces* the certainty of having a good strategy to achieve the desired final state.

PLAYING WITHOUT RISK

Not all problems can be resolved with this method. For example, let's consider self-flying drones (NASA has been working on a prototype since the early 2000s). It is not reasonable to build a million drones, send them up in the air, and watch them crash to the ground over and over until we eventually figure out the right parameters. Only the Shadoks[1] would use this method. Computer scientists use simulation instead. However, the algorithm only learns how to behave in simulation. There's no guarantee it will work correctly in real life!

Thus, reinforcement learning can only be used for applications that simulate no-risk operations, as is the case with Go. Using this reinforcement learning technique, AlphaGo engineers succeeded at gradually training the two neural networks: all you have to do is have the computer play against itself. This produces billions of matches that allow the neural network to gradually reinforce its confidence to make decisions, whether it be to reject a move (the "network policy") or to stop the game (the "value network").

GIVE ME THE DATA!

The same reinforcement learning technique can be used to teach AlphaGo to play chess and checkers. In 24 hours, the system calculates the best action sequences and becomes unbeatable.

---

1   The Shadoks are silly animated cartoon characters, very popular in France in the 1960s, that keep failing in building all sorts of machines.

These two learning methods, neural networks and supervised learning, rely on a similar idea. This consists of writing an algorithm to adjust a set of parameters in a calculation that allows the machine to make a decision. This adjustment is achieved by exploring a new space: the *parameter* space. When your system receives an input, it calculates an output based on the input's parameters. The learning algorithm is going to explore this set of possible parameters, which is gigantic, to find values that allow it to obtain a good response ... in most cases.

When the problem to be resolved is easy to simulate, as is the case with Go (losing a match isn't a big deal), the reinforcement learning techniques allow this parameter space to be explored in a clever manner. In other cases, the system has to be given examples. A lot of examples, often in un-fathomable quantities. Thus, to be trained correctly, Rosenblatt's perceptron needs more triangles than a geometry teacher sees in a lifetime!

However, in the era of "big data," satisfying this need is no longer a problem. Artificial intelligence research laboratories build gigantic databases to train their algorithms. And if Google is the best at image recognition, natural language comprehension, and Go, it's because their engineers have the ability to collect data from our internet searches, our telephones, our videos, and our emails.

Certain domains continue to elude these learning-based techniques because labeled data aren't available. Consequently, artificial intelligence researchers develop techniques using symbolic models, like the ones we saw in the early chapters of this book, to generate synthetic data based on computer models of a given domain. These synthetic data capture a degree of human expertise.

These very promising approaches have been in development for some years and have a bright future ahead.

And, just like that, computers now regularly outperform us. It's going to take some time to get used to – even though we are solely responsible for this!

Strong AI

Understanding AI's Limitations

15

Our journey in the land of AI will soon be at an end. What we've seen so far is a general overview of some problems and some methods used to solve them. There are still many other techniques we haven't covered. Most approaches used in AI require many years of study before they can be fully mastered.

Despite this, you get the general idea: humans build machines that perform calculations, transform data, and resolve problems. However, these machines are not truly intelligent. They apply a procedure created by a human being.

So, is it possible to build machines that are truly intelligent? Machines capable of resolving all kinds of problems without human intervention? Machines capable of learning and developing as children do? Machines aware of the world around them? Machines that can feel emotion, form a society, and build a future together? Some people, such as French researcher Jacquets Pitrat, firmly believe so.

We might as well say it now: this is a tricky topic that has more in common with science fiction than with actual science. It goes without saying, the dream of a truly intelligent machine made in humans' image has been an extraordinary driving force behind all of the developments in artificial intelligence since the 1950s. Currently, however, there is nothing to suggest it is possible to build such a machine.

STRONG AI

The artificial intelligence techniques we have seen throughout this book were systematically developed for a very specific goal: to effectively resolve a given problem. They do not attempt to imitate humans in every domain.

In the 1970s, the philosopher John Searle hypothesized that the human brain could behave like a computer, a machine that processes information. Subsequently, he used *Strong AI* to refer to artificial intelligence capable of perfectly imitating the human brain. The term stuck.

Today, we use the term *weak AI* to refer to AI techniques that allow us to resolve specific problems, such as finding a path for a GPS, recognizing cute kittens in pictures, or winning at Go. It's a bit of a misnomer: these computer programs systematically surpass human ability; what's more, that's what they were designed for. You might agree that it's a little presumptuous to say they're "weak." Human pride knows no limits.

Strong AI, on the other hand, refers to artificial systems capable of thinking like a human being. As a result, researchers distinguish between two very different problems: building a general AI and building an artificial consciousness. Let's take a look at these two notions.

GENERAL INTELLIGENCE

The idea of developing a machine capable of adapting to any type of situation and resolving a wide variety of problems like a human wasn't thought up yesterday. It is certainly the main driving factor behind countless artificial intelligence research projects. Although a great number of researchers have given up on this utopia, many still hear its call.

General artificial intelligence brings together researchers who develop AI in this direction. Their objective is to develop an artifact capable of performing any intellectual task a human can accomplish. Consequently, they have developed a certain number of tests that a general AI must be able to pass, such earning a university degree, passing the Turing test, or preparing a coffee in an unfamiliar house.

The difficulty is that the *same program* must perform all of these tasks, whereas a weak AI uses one program for each set of problems. Attention! This doesn't mean that a so-called weak artificial intelligence cannot be applied to several problems. A primary objective in AI and any computer science research project is to develop algorithms that are as generic as possible, regardless of the method used. Thus, the minimax algorithm is the same for chess as it is for Reversi. The only thing that changes is the heuristic function. A deep neural network will always have the same structure. The only things that change are the number of neurons in each layer and the learning examples provided to the machine.

However, this isn't the only difference between a general AI and a weak AI. It is absolutely impossible to apply a weak AI algorithm directly to a problem that it hasn't been designed for. If you have any doubts, just try resolving the traveling salesman problem using the AlphaGo neural networks!

ARTIFICIAL CONSCIOUSNESS

The second problem that research into strong AI looks at is how to build an artificial consciousness – a machine that would be conscious of physical or immaterial elements in addition to the data it manipulates, a machine that would be aware that it's a machine.

Researchers in this domain raise questions that are entirely different from the ones we've studied so far. First and foremost, scientists attempt to define what consciousness is and what a machine would have to produce to allow us to say that it is conscious. This is an open question and it relates to subjects that draw on philosophy and neuroscience just as much as artificial intelligence.

What would the neural activity of such a consciousness look like? Could this phenomenon be reproduced in a machine? Under what conditions? An underlying idea in artificial consciousness research is that a machine capable of simulating a human brain would be able to create a consciousness, but philosophers are skeptical of this purely mechanical vision of consciousness. Numerous researchers believe the body also plays a fundamental role in consciousness, just as it does with emotion.

AN UNCERTAIN FUTURE

Is a general artificial intelligence or an artificial consciousness around the corner? It is impossible to give a purely scientific answer devoid of opinions, beliefs, and ideologies. Many researchers do not "believe" it is possible to create a strong AI with our current means. However, no one can fully prove that it is impossible, and several businesses are investing in this type of research.

History has shown that the future of artificial intelligence is difficult to predict. Certain things that looked on the verge of being resolved in the 1980s, such as diagnosing illness, have proven to be terribly difficult. On the other hand, certain objectives that seemed impossible to achieve, such as Go, have clearly become much easier. In the 1970s, no one would have bet a dime on neural networks. Today, they're a major focus of attention.

While it may not lead to any concrete results, strong AI research certainly contributes to the knowledge we have about human intelligence. It would be foolish to dismiss it outright.

HOW FAR CAN WE GO?

In computer science, and AI in particular, making predictions is rather audacious. Many people have been wrong about AI, and many will continue to be wrong about what AI has in store for us. Even if we remain within the well-controlled domain of so-called weak AI, it will surely open new doors for society. Machines are increasingly capable of accomplishing what might have seemed impossible to automate just a few years ago. And, as you might expect, this evolution is raising concerns.

Is There Cause for Concern? **16**

Understanding That AI Can Be Misused

Artificial intelligence is a powerful tool. It allows machines to accomplish a considerable number of tasks that, until very recently, could only be accomplished by humans. Computer applications written by AI algorithms also allow us to process data faster, whether it's for making decisions, travel, finding information, controlling industrial machinery, or just playing games.

Quite often, AI's capacities greatly surpass those of humans. This inevitably leads us to consider the place of machines in our societies and our relationship with them. Will they eventually take control of our societies and destroy or enslave humanity, as in sci-fi movies?

AUTONOMOUS SYSTEMS?

Not with what we know at present. "Weak" AIs have no ability to create or imagine. For this reason, they are unable to go beyond the framework programmed for them. Even when we talk of "autonomous vehicles," or self-driving cars, the autonomy in question is limited to what the machine has been designed for. The car will not "turn" against its driver unless it has been programmed to do so.

Isaac Asimov illustrated this situation very well in his *Robot* series. The characters are machines that obey the stories' three laws of robotics, the first being to never injure a human being. Throughout the stories, we discover that the robots can adopt very strange behaviors, and even cause accidents. However, never do the robots violate the laws of robotics. No robot ever injures a human or allows a human to come to harm.

Such "autonomous" systems are becoming more and more common, and we can see they are relatively harmless indeed... as long as we use them correctly! This is exactly like other tools we use on a daily basis. Lots of people get hurt each year with a screwdriver, a drill, or a saw, occasionally because of a defect, but more often because of misuse. However, these objects are not conspiring against us. This only happens in horror movies.

As for strong AI that would be self-aware and determined to break free of our control, there is nothing preventing anyone from claiming it may exist one day. Artificial intelligence, the generalization of "big data," and recent advancements in deep learning are undeniable leaps in technology. Yet they don't bring us any closer to artificial consciousness. They are still just tools at the service of humans.

MISUSE OF AI

The main danger with AI today is how these technologies can be used. Let's consider search engines like Google and Baidu, for example. They allow anyone to access any information contained in the giant library that is the internet. But these same algorithms can be used to automatically filter the content you access. This is also what Facebook does when it calculates the information that might interest you based on your profile.

This wouldn't be so bad if social networks hadn't become the main source of information for persons aged 18–24. What's to prevent an information dictatorship controlled by tech giants or totalitarian states from deciding overnight what we can or cannot know?

It's the same thing for almost every AI technique. What's to prevent the image recognition or automated decision-making technology currently used to develop self-driving cars from one day being used to develop killer robots capable of identifying their targets and eliminating them without human intervention? Many researchers including Stuart Russell have already raised alarms about the malicious use of AI technology.

And they are right: it is urgent that we understand what AI algorithms are capable of so that we, humans, can decide what we choose to consider acceptable. We need to understand which AI uses we wish to prevent and how to guard against them. It is up to us to develop algorithms that do not violate these ethical rules – not even by accident.

AI SERVING PEOPLE

Computer programmers, users, and politicians are all responsible for how AI will be used: above all, this is a legislation and ethics problem. It is important that we guard against potential malicious acts guided by political, religious, or economic ideologies. More than anything else, we mustn't forget that artificial intelligence is there to help us, not to imprison us!

Every day we use dozens of AI algorithms for our greater good, from braking systems on cranes and subway trains to information search tools that allow us to find an address, prepare a school presentation, or find our favorite song. These uses are often so insignificant that we don't even realize we're using AI.

Without a doubt, AI programs will one day be used to commit crime. In fact, this has most likely already happened, even if only in the domain of cybersecurity: hackers use programs to accomplish tasks that, to them, are "intelligent."

This is AI. As a result, cybersecurity specialists develop AI algorithms to guard against this. They detect server intrusions with the help of machine learning and isolate infected network computers using multi-agent systems.

We mustn't aim at the wrong target: AI can be misused, but it doesn't act on its own. It has been programmed by a programmer for a specific purpose. It has been used by a human with an objective in mind. In every instance, it's a human who is ultimately responsible for how it is used.

One mustn't confuse a tool and its use. An AI can be misused. An AI cannot, however, become malicious spontaneously, at least not based on today's scientific knowledge.

EXPLAINING, ALWAYS EXPLAINING

One of the biggest issues with artificial intelligence is helping users understand what "intelligent" machines do.

Self-driving cars will be on our roads soon. There will be accidents – it's inevitable. When this happens, we will need to determine why the system made an error, whether it was used correctly, and in which circumstances this error might reoccur. This is why more and more researchers are taking an interest in developing "explainable" AI algorithms.

Today, it's very hard to explain an AI's decision, whether it's to play chess, to recognize a cat picture on the internet, or to drive a vehicle. The inputs are transformed into numbers, which are fed into tons of calculations. Therefore, it is not easy to determine which part of a calculation is responsible for a certain aspect of a decision.

Some researchers attempt to write mathematical proofs to assess, or not, certain properties of a system. Others attempt to identify which properties have most contributed to the decision the system has made, such as recognizing the cat's ears in the triangle example. The objective is to produce an output that is intelligible enough for us to subsequently explain it based on the program's initial settings.

This is a difficult subject, and I don't think we can currently say this issue has been resolved. However, there is no doubt that in the coming years many AI algorithms will be developed that will be able to explain what information was used to form the basis of a decision. There may even be, as Stuart Russell proposes, algorithms capable of automatically detecting when they are about to make an error.

To Infinity and Beyond! **17**

Understanding That AI Has Good Days Ahead

Artificial intelligence is transforming our world. It is giving us machines capable of accomplishing tasks that, for a long time, we believed only humans are capable of. And while machines do not "think" in the same way humans do, AI's progress is nothing short of incredible. This is why it inspires us to dream big.

WHERE ARE WE GOING?

So many amazing things have been accomplished thanks to artificial intelligence. In light of so many advancements, it's hard to know what comes next. So many challenges remain. Some of them appear to be within reach, others light years away, no matter how clever the person predicting what AI's next success stories will be.

Today, we know how to program machines that can easily win at Go or poker; however, AI is unable to beat humans at video games like League *of Legends* or StarCraft, despite the ability of a computer to give thousands of commands per second. Programs can find information in a medical encyclopedia faster than any human, and they can isolate genes, but they are still unable to make a medical diagnosis. Algorithms can find a path by accounting for traffic jams and will soon be able to navigate a car through town, but they are unable to explain why.

The reasons are no mystery: the partial perception of the situation, the consideration of temporality, and the diversity of possible solutions. For now, though, AI researchers are unable to get around these obstacles. But each method contributes some progress, each algorithm has its victories. Gradually, our knowledge grows and we are able to build ever more impressive machines.

DOING LIKE HUMANS?

Some people think machines will soon be able to automatically correct philosophy essays or hire people using a resume and a video. I don't think so. The technical limitations aside, the human factor in these tasks is so subtle that it would be difficult to capture it in an AI system. Humans would quickly find out how to trick such systems using "antagonist data."

Perhaps one day we will discover that artificial intelligence cannot do everything. That there are some things that elude computer data processing. Or perhaps not. As more discoveries are made, will we ultimately succeed at imitating humans flawlessly, including their abilities to err, to behave foolishly, and, despite everything, to strive tirelessly to improve?

Artificial intelligence algorithm research teaches us a lot about humans. We often attempt to understand how humans perform tasks in order to build machines that can do it better. Before that, many researchers also attempted to build machines to understand man or nature: the neural networks, to cite one example, were invented to study the theories proposed by psychologists in the 1950s. Thus, the progress with machines is also raising questions about humans' thinking mechanisms.

In the end, isn't this the greatest gift AI could give us: the ability to better understand ourselves?

Even More AI!

Sadly, it impossible to talk about all of the AI techniques invented since 1956: we would need another 15 chapters! My goal in this book was not to cover the entire spectrum of artificial intelligence research, but to simply help you understand how AI works, what its limitations are, and how, over the course of history, we have progressed in developing machines that are increasingly capable of performing tasks that require "intelligence."

AI techniques are used in numerous domains even if we are not entirely aware of it. In fact, we use AI every day. At this point, I think it is fair to credit some famous researchers for their discoveries in artificial intelligence that I did not have the opportunity to speak about in the preceding chapters. Dozens of names deserving mention come to mind. I've chosen the most famous ones below.

JAMES ALLEN

James Allen is a researcher who isn't content to make only one major contribution to artificial intelligence. James Allen is famous for his temporal reasoning model that allows computers to reason about notions such as "before," "after," "at the same time," and "during." He has also helped develop several natural language dialogue systems that combine knowledge representation techniques in *symbolic* artificial intelligence with *statistical learning.*[1]

---

1 These terms are explained in chapters 10 and 11.

JOHN MCCARTHY AND PATRICK HAYES

In the 1960s, John McCarthy and Patrick Hayes laid the foundations for an entire field of symbolic artificial intelligence. They proposed a model for reasoning about actions and changes: things are no longer always true or always false. In so doing, they paved the way for *cognitive robotics* research – in other words, building robots capable of reasoning about how to behave in the world. This isn't quite artificial consciousness, however, because the machines simply apply rules of mathematical logic within a framework whose bounds have been set by humans. However, the systems developed using this technology are able to adapt to different situations by using *logical reasoning*.

RICHARD FIKES, NILS NILSSON, AND DREW MCDERMOTT

In artificial intelligence, planning is the branch that concerns itself with calculating the right sequence of operations to achieve a goal. The STRIPS model, invented by Richard Fikes and Nils Nilsson in 1971, helps to describe actions, to describe a goal using a well-formed formula, and to calculate actions to achieve this goal. Planning algorithms allow systems to autonomously adapt to different situations. They are used in numerous industrial systems… and even in the Hubble telescope!

Other models and algorithms have been proposed since then, such as Drew McDermott's PDDL language, or the hierarchical task network (HTN) models proposed in the early 2000s. Each year brings new advancements that allow machines to resolve increasingly complex problems and come one step closer to human reasoning.

CHRISTOPHER WATKINS, LESLIE KAELBLING, AND MICHAEL LITTMAN

The best-known *reinforcement learning* algorithm[2] is certainly the *Q-learning* algorithm proposed by Christopher Watkins in 1989. It is taught in artificial

---

2   This term is explained in chapter 14.

intelligence courses at universities all around the world. In 2014, a Google team used a *Q-learning* variation to develop an AI program capable of playing Atari's 1980s video game Breakout just as well as a human. Today, this type of algorithm is commonly used in industrial robots to grab objects.

In the 1990s, Leslie Kaelbling and Michael Littman took a model from another branch of computer science and adapted it to artificial intelligence: *the partially observable Markov decision process.* When combined with reinforcement learning methods, this tool allows machines to calculate optimal solutions for *planning* problems by considering eventual uncertainties in the information transmitted by the sensors. These planning algorithms are widely used in robots, in particular the robots NASA sends to Mars.

ROBERT KOWALSKI, ALAIN COLMERAUER, AND PHILIPPE ROUSSEL

In 1972, French researchers Alain Colmerauer and Philippe Roussel proposed the first programming language based entirely on logic, drawing inspiration from the work of Britain's Robert Kowalski. The language, Programming Logic (Prolog), would become a big hit in the artificial intelligence community. Widely used in automated systems, including the Paris subway, it is still taught at universities today. More recently Prolog was used in programming Watson, IBM's artificial intelligence computer system that beat all of Jeopardy's champions hands down.

EDWARD FEIGENBAUM

Edward Feigenbaum is considered the "father" of *expert systems.* These computer programs use logical reasoning to emulate the decision-making ability of human experts: chemists, doctors, security experts, and so forth.

Expert systems are still used today in large industrial infrastructures.

LOTFI ZADEH

Lotfi Zadeh is the inventor of *fuzzy logic.* Behind the odd name lies a very particular computing model that allows computers to reason about facts that are more or less true, such as "it's nice outside," "it rained a lot," or "the car is going fast." Artificial intelligence programs using fuzzy logic, with reasoning rules written by humans, have been used in all kinds of intelligent control systems since the 1980s: cranes, elevators, meteorology, air traffic control, and many others.

ALLEN NEWELL

Allen Newell is an artificial intelligence pioneer who was present at the Dartmouth workshop in 1956. In the 1980s, his work led to *cognitive architectures*, the most famous of which are undoubtedly John Laird's Soar and John Anderson's ACT-R. These are computer models that allow a computer to simulate human behavior in a specific situation. They are used in video games to provide nonplaying characters with more realistic reactions when confronted with players' decisions.

Human behavior simulation is increasingly used in decision-making support systems: it helps leaders, politicians, and business owners analyze and understand complex phenomena such as opinion dynamics in politics, energy consumption, or even urban space occupation.

JUDEA PEARL

Judea Pearl is the inventor of *Bayesian networks*. This is an artificial intelligence method that combines symbolic reasoning (the objects manipulated by the Bayesian network can be interpreted by humans) with statistical learning (these objects are related by the probabilities that may be learned from the data). Bayesian networks are used in numerous applications to represent relationships between facts, in particular in bioinformatics, information research, and decision-making support systems.

RICHARD RICHENS, JOHN SOWA, AND RONALD BRACHMAN

In symbolic AI, machines attempt to reason about the knowledge contained in data, just like we do. For this, *knowledge representation* models are developed to program computers with the knowledge they are meant to manipulate. The first general knowledge representation model was proposed by Richard Richens in 1956. He dubbed this model a *semantic network*, and it led to many other advancements in artificial intelligence such as John Sowa's conceptual graphs, proposed in 1976, and Ronald Brachman's *description logics*,[3] proposed in the mid 1980s.

DEBORAH MCGUINNESS

In the early 2000s, symbolic AI specialists invented the *semantic web*. This can be defined as a network of web pages written especially for computers using well-formed formulas so they can read internet content just as humans can. The semantic web uses computer models based on description logics and the OWL language invented by Deborah McGuinness and her colleagues.

Nowadays, computers exchange "semantic" information on the internet to respond to our needs. For example, when you search for a celebrity using a search engine, you obtain a text box showing the person's age, height, and general information. This information is taken from data in the semantic web.

---

3 This term is explained in chapter 10.

Acknowledgments

I first discovered artificial intelligence at college. And I never looked back. This book would not have been possible without all the extraordinary people I've met. I owe them so much. Sylvie was the first person to introduce me to AI. Jean-Paul guided me during my first forays into research. Fred, Seb, Sophie, Alex, my colleagues, and friends, helped me find the right words to make difficult AI notions easier to understand. In the last 20 years, I've met so many passionate researchers, fascinating colleagues, and friends that it is impossible for me to name them all here. You know who you are. Know that this is also your book.

I would like to give special thanks to Monique and Alain, who have led hundreds of students down the path of AI and without whom I would not have met my favorite artist, Lizete. I am grateful for the opportunity to follow in your steps and guide young people down a promising path in AI.

This book owes much to my wife and children, who patiently reviewed each chapter and each contributed their grain of sand. They know I love them more than anything. Thank you to my friends, Céline and Paul, for their support and their readings. Thank you, Lizete, for your patience and unwaveringly good humor! Writing for the general public is not easy when you spend most of your days speaking with experts. Without all of your advice, this book would never have seen the light of day.

I would also like to thank Emmanuel, who helped me with the arduous process of publishing this book, Kaska, who supported me in the whole process, as well as my publishers, Ms. Baud and Ms. Cohen, for putting their trust in us.

Finally, thank you to everyone who has made, and continues to make, history in AI. I am especially thinking of the many researchers working behind the scenes who keep the small AI community ticking. The AI society worldwide have had to weather storms, buck trends, and withstand changing opinion for 20 years to ensure AI's survival. Rarely do they reap the fruits of their labor, and yet without them it would not be possible to write this book about AI today.

N. Sabouret

In the beginning, it was just a trio of computer scientists who posted about big things on an internet blog. The blog continued on its merry way in the hallways and break rooms at the LIP6 for 11 more years until Nicolas got back in touch with me about this incredible collaboration. It is with deep emotion that I would like to thank my former blog sidekicks: Nicolas Stefanovitch, a.k.a. Zak Galou, and Thibault Opsomer, a.k.a. Glou. I would also like to acknowledge my art companions: all the members of the CMIJ, of the AllfanartsV2 forum, and of BD&Paillettes. Thanks to François for his help with logistics (dishes, cooking, and babysitting Renzo), which allowed me to put on my artist apron every night in December 2018. Finally, thank you, Nicolas, for your enthusiasm and availability for this beautiful colla-boration. I am happy I was able to fulfill my lifelong dream of helping others learn about science through my artwork.

L. De Assis

They made AI

but they were not alone...